Safety in the Chemistry and Biochemistry Laboratory

Related Books from VCH

W. Jeffrey Hurst (editor)
Automation in the Laboratory

Hans H. Rump and Helmut Krist
*Laboratory Manual for the Examination
of Water, Waste Water, and Soil*
Second Edition

Safety in the Chemistry and Biochemistry Laboratory

André Picot and Philippe Grenouillet

Andrew T. Prokopetz and Douglas B. Walters
Editors of the English-language Edition

VCH

André Picot
Research Director
Chemistry Institute, Natural Products
CNRS
91198 Gif-sur-Yvette
France

Philippe Grenouillet
General Electric Plastics ABS,
Paris, France

Translator:
Robert H. Dodd
CNRS
91198 Gif-Sur-Yvette
France

This book is printed on acid-free paper. ∞

Library of Congress Cataloging-in-Publication Data
Picot, André.
    [Sécurité en laboratoire de chimie et de biochimie. English]
    Safety in the chemistry and biochemistry laboratory / André Picot
and Philippe Grenouillet: Andrew T. Prokopetz and Douglas B.
Walters, editors of the English-language ed.: [translator, Robert
H. Dodd].
        p.   cm.
    Includes bibliographical references and index.
    ISBN 1-56081-040-8 (acid-free)
    1. Chemical laboratories--Safety measures. I. Grenouillet,
Philippe. II. Prokopetz, Andrew T. III. Walters, Douglas B.
IV. Title.
QD63.5.P5313   1994
542'.1'0289--dc20

94-38824
CIP

© 1995 VCH Publishers, Inc.

Originally published under the title: "La Sécurité en Laboratoire de Chimie et de Biochimie,"
Second Edition, by Lavoisier TEC&DOC, 11 rue Lavoisier, 75384 Paris, France

Printed in the United States of America

ISBN 1-56081-040-8 VCH Publishers

Printing History:
10 9 8 7 6 5 4 3 2 1

Published jointly by

VCH Publishers, Inc.   VCH Verlagsgesellschaft mbH   VCH Publishers (UK) Ltd.
220 East 23rd Street   P.O. Box 10 11 61   8 Wellington Court
New York, New York 10010   69451 Weinheim, Germany   Cambridge CB1 1HZ
                                                            United Kingdom

Foreword

In France, a common saying is "safety is a daily affair and is everyone's concern." In fact, my contacts with public and private research in Great Britain and the United States, as well as in France, have taught me that mastering safety, whether in a research laboratory, a production unit, or even in industry, is first of all a matter of attitude. In the case of chemical substances, as with many other sources of risk (physical or biological), the development of bad habits must especially be avoided. Taking risks for granted is without doubt the number one enemy of safety.

It is indispensable that, from the onset, all prevention strategies integrate training and information for all, whether administrative or scientific personnel. The material means assuring these strategies must also be made available.

Chemical substances are not solely the property of chemists; biologists, physicists, electronics specialists, and even astronauts are continually in contact with such substances. Chemistry is universal and, though contributing to society's well-being, it must be adequately controlled. This control must always be improved so that both the quality of research and safety in the work environment benefit. Safety cannot be dissociated from high-quality work.

Throughout this work by André Picot and Philippe Grenouillet, I have been made aware of basic ideas on safety which no responsible scientist can ignore. It is the duty of all laboratory heads to properly teach the basic rules of safety to personnel and to provide them with the safest possible working conditions.

This excellent treatise on safety in the chemistry and biochemistry laboratory is, in my opinion, original in many respects.

Concerning the authors, André Picot, an organic chemist trained in the

v

Roussel-Uclaf laboratories, has oriented his research toward toxicochemistry, a field at the interface of chemistry and biology, at the Institut de Chimie des Substances Naturelles of the CNRS. Philippe Grenouillet, who studied at the University of Bordeaux, has been safety engineer for the CNRS for more than 10 years. He is now an acknowledged specialist concerning safety problems that can arise in chemistry and biology laboratories. The complementarity between these two authors is evident throughout the book, in which their scientific rigor and pragmatism is brought to the forefront.

I have found Chapter 2, dealing with the destruction of chemical substances, particularly interesting and valuable. The reactions and the procedures are clearly and precisely described. These methods will be of great utility to practising chemists.

In conclusion, it is my sincere hope that this important text will enjoy a long career befitting the considerable amount of information it contains. It is too often forgotten that personnel are often confronted with risks but that these risks can be minimized if proper precautions are taken. Among these precautions, information and training are paramount; this book by André Picot and Philippe Grenouillet is unquestionably a valuable help in this quest.

D. H. R. Barton
Nobel Prize 1969
Professor Chemistry
Texas A&M University
College Station, Texas

Preface

It is always difficult to classify by order of priority the factors necessary in diminishing risks. Most of these factors cannot, in fact, be dissociated. Moreover, priorities are different depending on the function and responsibilities of personnel as well as on the general awareness of risk and prevention in a laboratory or institute.

It appears to us that, in most research laboratories, the dissemination of information concerning chemical substances as well as the identification of risks involved in handling them, remains a priority. This information must allow increased awareness of danger and permit the medical corps, safety engineers, and committees to work more efficiently. This includes adequate knowledge concerning dangerous substances and intermediates that may be formed and which can determine the stability and toxicity of mixtures; the risks inherent to certain reactions and manipulations; the quality of the technical environment and the organization of work. Various methods allow analysis of professional risks, thereby allowing detection and minimization of these risks. The information that we have collected in this book should help in this type of analysis.

Once dangerous situations have been identified, there remains the difficult task of creating a professional environment and of formulating individual and collective rules which can reduce these risks. Regulations and techniques of protection are efficient tools. Those which we present in this book may help to organize a safe working environment.

We have also tried to deal with particular problems presented by some substances. This is why we have illustrated general techniques with a large number of specific examples that can easily be accessed using the index that we have attempted to make as complete as possible.

When an incident or an accident occurs, it must be dealt with in such a way that the situation is not made worse. This is another aspect of prevention. During the course of descriptions of dangers associated with certain chemical families, we have attempted to give general rules of conduct in case of mishap. However, this text is not meant to be used as an emergency handbook.

An important place has been given to the treatment of wastes and aged chemicals. Experience has in fact revealed to us that mismanagement of stocks and wastes is often a cause of accidents. This subject merits considerable attention in the design of any new research facility. The "recipes" that we present are nothing more than one of the aspects of the organization necessary wherever a large number of diverse substances are employed, regardless of quantities.

The nature of substances that are handled, as well as techniques, change with time. Knowledge concerning the toxic properties of chemicals also evolves. While it is obviously not possible to cite all known substances, the choices we have made are based on our experience in research laboratories and on qustions that are asked during the course of training sessions. With this in mind, observations and comments by readers will be extremely valuable in completing this work in the future.

Philippe Grenouillet
André Picot

Acknowledgments

We are particularly grateful to Professor Derek Barton who was kind enough to preface this book. We would also like to extend our thanks to Drs. Ourisson and Potier, members of the French Academy of Science and successively directors of the Institut de Chimie des Substances Naturelles (ICSN) of the CNRS, who encouraged and supported us throughout our project.

We are indebted to Drs. Cherest and Millet and Mr. Cosson of the ICSN for having proofread and corrected the manuscript.

We would also like to thank all our colleagues who offered their help and advice, especially members of the ICSN safety committee, Mr. Brendel, safety engineer at the CNRS, Mr. Castegnaro, research scientist at the International Cancer Research Center, Mr. Gaignault, chemical engineer at Roussel-Uclaf, Mr. Gaillardin, safety engineer at INSERM, Mrs. Mousel, staff medical doctor at the Pasteur Institute, Mr. Pezerat, CNRS research scientist at the University of Paris 7, Mr. Rogelet, safety engineer at the CNRS, Mr. Rollin, research scientist at the Pasteur Institute, Mr. Simonnet, head of teaching at the Institut National des Sciences et Techniques Nucléaires, Mr. Tessier, safety inspector at the CNRS, Mr. Toromanoff, chemical engineer at Roussel-Uclaf, and Mr. Zerbib, engineer at the CEA.

Contents

PART 2. Other Risks in the Chemistry Laboratory

3. Biological Risks   243

4. Laboratory Risks Associated with Nonionizing Radiation   255

1

CHEMICAL RISKS

Introduction — General Rules

The person or committee responsible for hygiene and safety must be vigilant in adapting the following general rules to the specificities of the laboratory by first defining them (instructions for the use of hazardous substances such as genotoxic substances, regulations for off-hours work and for storage areas, periodic audits, names and telephone numbers of first-aid volunteers, etc.), by making them known (posters, teaching sessions), and in enforcing them. All persons implicated must participate in the elaboration of general or particular regulations; this is the best guarantee of their effectiveness.

1.1 Organization of the Working Environment

–The organization of laboratories (distribution of workbenches, equipment, ventilation systems, operating regulations) must be the subject of rigorous study before their construction or modification.

–Experiments, especially those presenting major risks (hydrogenation, ozonation, distillation) or performed in isolated quarters, must never be conducted alone.

–Hazardous operations must not be undertaken outside normal working hours. In case of accidents at night and during holidays, the delay before help can arrive is always longer.

–Before engaging in hazardous work, colleagues should be informed. These experiments must be continuously monitored and the absence of the experimenter, even for a brief period, should be avoided.

–Experiments must be undertaken only on clean and unencumbered work-benches or fume-hoods.

–The number of gas outlets should be limited; one per laboratory is usually sufficient. The main cutoff valves for water, gas, and electricity must be well indicated, easily accessible, and regularly inspected.

–Flammable substances must be stored in ventilated metal locked storage facilities.

–The total amount of flammable substances (solvents) stored in the laboratory should be limited.

–The ventilation systems must be regularly inspected and maintained. The use of air velocity meters and smoke tubes can assist in these inspections.

–Flammable materials must not be stored or handled close to an open flame or heating device.

–Pipetting by mouth must be forbidden, even for supposedly nontoxic substances.

–The storage within the laboratory of large quantities of solvents and of reagents prone to spontaneous decomposition at ambient temperature is to be avoided.

–Corrosive products (strong acids and bases, etc.) as well as toxic and flammable substances should not be stored on shelves located above the workbench.

–The prolonged storage of substances in direct sunlight should be avoided.

–Special locked storage facilities must be used for substances reacting violently with water (alkali metals, etc.) and for toxic products (poisons, biologically active compounds, etc.).

–Refrigerators must be modified such that thermostats and incandescent bulbs are removed from the interior, if flammables are to be stored within.

–Inventories of the contents of refrigerators, freezers, and cold rooms must be performed regularly (at least annually). These must be thawed to avoid condensation of toxic, flammable, or unstable products onto the cooling mechanisms.

–Special safety precautions must be taken whenever the usual working conditions are modified, notably when external servicing companies are called in.

–Never dispose of waste down the sink, especially the following:

· Compounds reacting violently with water (alkali metals, organometallics, hydrides, etc.)
· Toxic compounds (phenol, cyanides, certain heavy metal salts: mercury, lead, thallium, chromium, cadmium, etc.)
· Flammable substances (solvents, etc.)
· Foul-smelling substances (mercaptans, etc.)
· Lachrimators (acyl halides, etc.)
· Substances not easily biodegraded (polyhalogenated derivatives, etc.)

–For reasons of hygiene, eating and drinking in laboratories, especially out of laboratory glassware, should be prohibited. Smoking in the work area, especially in the course of manipulations, must be prohibited.

1.2 Emergency Interventions

–Have on hand all the following useful telephone numbers (preferably in an Emergency Procedures Contingency Plan):

• emergency phone number for fire, police, and hazmat team.
• first-aid officer
• infirmary and medical doctor
• the laboratory safety officer
• hygiene and safety services
• poison center

–Participate in fire-fighting classes.
–Always keep on hand inert absorbants (supercel, etc.) in case of spills. It is judicious to maintain a good supply in the laboratory stores.
–In cleaning up mercury spills in the laboratory, use the method best suited to the situation. (Large quantity spills may require initiating the lab's contingency plan and/or contacting your local hazmat team. Smaller spills may utilize the techniques proposed starting in Section 2.23).

1.3 Individual Protection

1.3.1 Protection of Eyes

Projections and explosions generally occur without warning and, among laboratory accidents, those affecting the eyes are the most frequent and the most serious.

The wearing of safety goggles at all times in the vicinity of a laboratory operation is imperative, even if manipulations are only being conducted by someone else.

Persons wearing corrective glasses must, when performing dangerous operations, wear supplementary protective goggles having lateral flaps or prescription safety glasses with side shields.

The wearing of contact lenses in the laboratory is not advised. Numerous volatile products (organic acids, halogenated derivatives) can dissolve in the ocular fluid in which the lens floats, provoking serious irritations (a phenomenon that is accentuated in the case of "soft" lenses).

The individual choice of protective goggles is important (comfort, esthetics, type of danger). Several models must be proposed both to the personnel and to new arrivals, visitors, temporary workers, and so on. A small supply of goggles should be kept on hand, especially in view of their relative ease of damage.

In the case of a particularly dangerous operation, it is recommended to work under a fume-hood equipped with a polycarbonate safety shield.

1.3.2 Protection of Hands

Caustic products (bromine, strong mineral acids, strong bases, powerful oxidizing agents) and compounds that easily penetrate skin (aromatic amines, nitro derivatives) must be handled with impermeable gloves (check permeability test results for the appropriate glove type). Moreover, certain chemical compounds (aromatic amines, hydrazines, etc.) can, by remaining impregnated in the gloves, be a source of poisoning (therefore use disposable gloves whenever possible).

Depending on their composition, gloves can be permeable to certain chemical products. The choice of gloves thus depends on the type of substance to be manipulated. It may even be necessary to wear two dissimilar, disposable types of gloves simultaneously (e.g., latex and vinyl for nitrosamines).

Even with the use of two dissimilar pairs of gloves, hands should be carefully washed after handling any hazardous materials, especially:

–toxic products (cyanides, arsenates, etc.)
–biologically active substances (estrogenic hormones, alkaloids, etc.)
–allergenic compounds (quinones, Ni^{2+} salts, etc.)

For noncorrosive toxic substances, the use of disposable gloves is recommended. The protection of hands cannot be provided by protective creams because these do not appear to have the necessary qualities of performance and tolerance. In particular, "liquid gloves" cannot replace real gloves in cases, for example, where there is a risk of contact with a hazardous material.

When inserting glass tubing or a thermometer into a stopper or releasing a frozen glass stopcock, hands should be protected with a thick rag or, better, with gloves (e.g., Kevlar-type gloves) designed for the handling of glass. These operations are, in fact, the most frequent cause of wounds, sometimes very serious (cut tendons, etc.).

1.3.3 Protection of the Respiratory Tract

When working with volatile, toxic substances (phosgene, hydrogen sulfide, chlorine, etc.), the use of respiratory protection is indispensable (if engineering controls or administrative controls such as changing chemicals is not practical). These respirators must not be used if the level of polluting elements in the air exceeds the protection factor for negative pressure respirators. The filter may have to be replaced after each use or when there is a change in the end of service life indicator.

In case of an accidental release of toxic gases, the area must first be evacuated before returning with an emergency response team equipped with Self-Contained Breathing Apparatus.

Knowledge of the location and operation of these respirators is indispensable. However, only trained employees who have been fit tested and medically approved can wear these properly selected respirators. Co-workers must be warned whenever volatile, toxic substances are to be employed.

1.3.4 Protection of the Body

A lab coat should always be worn and kept as clean as possible; contamination by way of soiled clothing is a nonnegligible risk. Disposable Tyvek lab coats and coveralls are preferable to cotton lab coats. Synthetic material whose resistance to fire and corrosive products is unknown must be avoided. It is preferable to work with closed footwear.

1.4 Ventilation

Laboratories must always be equipped with efficient means of aeration such that ambient air quality is guaranteed. Appropriate ventilation is an essential element in assuring laboratory safety. It is one of the best means of protecting individuals against pollutants regardless of whether these originate from the mere presence of humans (nonspecific pollutants) or from other sources (specific pollutants; e.g., gases, vapors, solid or liquid aerosols).

Ventilation may be divided into two types

- general ventilation
- local ventilation

The choice of ventilation type depends on:

- norms and regulations
- the maximum levels of exposure (MLE)
- the average levels of exposure (ALE)
- the risks of explosion or flammability
- the work to be performed
- the physicochemical properties of the substances to be employed
- the type of equipment to be used
- the number of people working

If two types of ventilation are to be installed, it is important to obtain a good balance between general and local ventilation.

In work areas, the ventilating equipment should not produce discomfort or nuisance due to the speed of circulating air, its temperature and humidity, its noise and vibrations, and so on. It is particularly important that the ventilating techniques employed do not significantly increase the background noise levels of the working environment.

1.4.1. General Ventilation

General ventilation allows the renewal of the air in a room:

$$R = \frac{D}{V}$$

where

> R = the rate of air renewal in a room
> D = the air flow in m^3 per hour
> V = the room's volume in m^3

The required rate of air renewal is more or less a function of the type of room (e.g., R = 5–12 air changes for a laboratory, R = 10–15 air changes for animal quarters) as well as the considerations cited earlier. Certain rooms may be positively or negatively pressurized depending on the type of protection sought. This can help to better protect products, the environment, or both. General ventilation does not allow complete elimination of pollutants but only dilutes them. If pollutants must be eliminated to a greater degree or completely, then local ventilation must be added.

1.4.2. Local Ventilation

Local ventilation allows better trapping of pollutants since it is employed closer to their zone of production. A certain number of possibilities is available: conventional fume-hoods, recycling fume-hoods with filtering of pollutants, flex duct, down-draft tables, glove boxes with filtration of incoming and outgoing air or replacement of air by an inert gas, and so on. Of these, the most commonly used in laboratories is the fume-hood.

In order to obtain the maximum efficiency and use from this type of ventilating equipment, the following "ten commandments" must be observed:

1. Enclose the area in which the pollutant is produced as much as possible.
2. Place the filtering apparatus as close as possible to the emitting source of pollution.
3. Install the ventilation equipment such that the operator is not placed between it and the source of pollutants.
4. Place the filtering apparatus such as to take advantage of the pollutant's natural movement.
5. Introduce sufficient air flow to trap the pollutants.
6. Ensure uniform air flow over the entire area from which pollutants must be trapped.
7. Compensate the outflow of air by a corresponding (or sometimes slightly lower) inflow of air (i.e., return air).
8. Avoid air currents and sensations of thermal discomfort.
9. Ensure that polluted air is expelled far from the entrance point of fresh air

and that all ducting is under negative pressure to avoid blowing the contaminant back into the building if there are leaks in the duct work.

10. Do not use fume-hoods whose dimensions are out of proportion with real needs or possibilities.

Regardless of the type of ventilation employed, the equipment must be maintained and inspected at least once a year. In the case of equipment having filters, prefilters, adsorption and/or absorption traps, it should be verified that these are able to trap both the type and quantity of expected pollutant. Their inspection should be made more frequently.

The choice of equipment to be used cannot be left to chance. Needs must first be defined that will allow both the most appropriate equipment to be selected as well as ensuring its optimal use depending on its limits and possibilities. Appropriate attire is also necessary during maintenance and inspection operations. Maintenance and inspection of the equipment must be performed regularly

1

The Handling of Chemical Substances

Before undertaking any new experimental work, acquaintance with the risks associated with the following elements is indispensable:

–the apparatus employed
–the physicochemical properties of the materials to be used, in particular, their stability alone or in mixtures
–the toxic properties of the substances to be utilized or which are likely to be formed.

These three subjects will be treated in the present chapter.

Though it is difficult to evaluate risks that are, a priori, unknown, it should be remembered that the safety of operations in which chemical products are to be used depends largely on the working conditions. Before undertaking any work, familiarization with the available means is necessary (the available work area, the usable apparatus, the disposition of each step of the operation, the nature of nearby operations, etc.). Vigilance before everyday, "no problem" reactions is advised.

1.1. Risks Associated with Apparatus

Before setting up equipment, the individual electric and glass parts must be checked.

All doubtful apparatus must be immediately inspected and repaired. Starred, cracked, or chipped glassware must be eliminated. Defective equipment must not be employed.

Equipment must be set up with a maximum of care. Only protected clamps must be used for ground-glass joints and glass elements in general. Ground-glass joints may be lubricated with an inert grease (silicons, etc.). Proper water circulation in condensers must be ensured and the strength and tightness of all flexible tubing verified.

Apparatus open to the atmosphere (i.e., not under vacuum) should be fitted with tubes containing dehydrating agents (e.g., calcium chloride).

1.1.1. Pressurized Apparatus

Operations performed under pressure present, in addition to the risks associated with the chemicals themselves, a serious danger of leaks due to overpressure or explosion. The apparatus utilized must be specifically designed for their purpose and carefully inspected (autoclaves, sterilizers, hydrogenators, pressurized chromatography columns, steel cylinders, sealed tubes, etc.).

Manipulations must never be performed by a person working alone and a protective screen must always be used.

1.1.1.1 Pressurized glassware

Starred, cracked, or deeply scored material must be eliminated. The various elements must be protected with safety shields. Particular attention must be paid during heating and cooling of pressurized equipment.

1.1.1.2 Sealed Tubes

The tubes to be used must be of excellent quality (perfectly clean, thick Pyrex glass). Tubes should be filled to one-third of their volume (never more than half) using a funnel, a pipette, or a syringe in order to avoid contact of the reaction mixture with the narrow part of the tube to be sealed. The contents of the tube, isolated from the external atmosphere (e.g., connected to a closed vacuum line), are cooled in liquid nitrogen before placing under vacuum. The narrow neck is then sealed with an acetylene torch in order to avoid the condensation of air inside the tube.

At the end of the reaction period, the tube and its protective envelope are removed from the oil bath or oven and, after cooling to room temperature, are left for a prolonged period in a liquid nitrogen bath. The extremity of the tube is lifted from the bath and a flame is applied to liberate any internal pressure. The neck can then be scored with a file and snapped open.

Remark: If this work is to be undertaken by a glassblower, it is indispensable that he be informed of the nature of the reaction mixture as well as of the possible dangers.

1.1.1.3 The Use of Compressed Gas Cylinders

All cylinders of compressed, liquified, or dissolved gases must be handled with caution, whether or not they are neutral, combustible, toxic, corrosive, or pyrophoric. The gas must be carefully identified before use by noting the color code on the neck of the cylinder (see Table 1.1) and the type of valve to be utilized.

All gas cylinders must be secured. The manometers must conform to the nature of the gas. The weight of filled cylinders must never exceed the sum of the tare and the maximum load. Quantities must be measured by weight differences rather than by reading off a manometer. The pressures indicated on

TYPE OF GAS	COLOR OF CYLINDER NECK
Acetylene	Dark brown
Hydrogen chloride	Light green + skull and cross-bones
Ammonia	Light green
Carbon dioxide (CO_2)	Grey
Sulfur dioxide (SO_2)	Light green + skull and cross-bones
Argon	Yellow
Nitrogen	Black
Deuterium	Red
Nitrogen dioxide (NO_2)	Light green + skull and cross-bones
Helium	Tan
Sulfur hexafluoride	Dark green
Hydrogen	Red
Hydrogen sulfide	Light green + skull and cross-bones
Liquid hydrocarbons	Pink
Krypton	Brown
Methane	Pink
Nitric oxide (NO)	Light green + skull and cross-bones
Neon	Brown
Carbon monoxide (CO)	Light green + skull and cross-bones
Ethylene oxide	Pink
Oxygen	White
Nitrous oxide (N_2O)	Blue
Xenon	Brown

Table 1.1   The conventional colors of gas cyclinders.

the manometer must never exceed the working pressure (WP), not to be confused with the hydraulic test pressure (TP; TP = 1.5 WP).

Remark: In the presence of carbon monoxide, never use nickel-containing manometers, tubing, or joints in order to avoid formation of nickel carbonyl (100 times more toxic than CO).*

The use of nickel alloy manometers in the presence of carbon monoxide L'Actualité Chimique. French Chemical Society, Paris (1984). pp. 67–68

Small cylinders (1 l) must be firmly supported and secured and should not be stored in the laboratory when not in use. Stored bottles must be regularly inspected (for the condition of cylinders, quantity of gas).

Cylinders containing dangerous gases (flammable, corrosive, toxic) must be stored secured in closed quarters away from buildings. If this rule cannot be observed (e.g., mobile chromatography installations using hydrogen as the mobile gas phase), the apparatus must be inspected before each use. In this last case, the use of a hydrogen generator is preferable.

> Cylinders must always be firmly anchored. A fallen cylinder can have very serious consequences if the working pressure is suddenly released into the atmosphere. The attachment of a manometer creates a weak spot in case of a fall, facilitating rupture.

Cylinders must always be transported on appropriately designed carts. The use of other techniques to move cylinders may lead to their falling or to bodily harm (crushed hands or feet, sprained backs).

All operations connected with the opening and closing of gas lines must be performed delicately. A stuck valve should be warmed with a rag dipped in hot water. If the valve still will not open, force, or worse, a hammer should never be used; the cylinder should simply be replaced. After use, the manometer pressure must be released, the flow valve left open, and the main cylinder valve closed without forcing. Valves must not be lubricated, especially those of oxygen cylinders.

> Toxic gas cylinders (phosgene, chlorine, hydrogen sulfide, arsine, phosphine, diborane, etc.) must only be opened by persons wearing an appropriate respirator or opened only in a ventilated enclosure. This operation must never be performed alone.

For such gases, the quality of the individual components of the apparatus is the primary assurance of safety. Additional safety measures must be implemented (flow control, automatic feed cuts in case of mishaps). Gas detectors should be employed (the use of an easily detected carrier gas, such as hydrogen, can be very effective).

Destruction of gases may be necessary (use of scrubbers). Trapping with water is preferred to thermal destruction. In the former case, dilution of the gas with nitrogen must be correctly performed. The effluents of such scrubbers must be checked, particularly when the manufacturer provides no guarantees. Compliance with local, state, and federal air regulations should also be checked.

Vacuum pump oil can dissolve certain toxic or unstable gases. Oil changes must be performed with caution (masks, ventilation).

1.1.2 Vacuum Equipment

Depending on the volume under vacuum, implosion provoked by thermal or mechanical shock can have effects comparable to those of an explosion. Allowing air to rapidly enter an apparatus under vacuum can lead to such

mechanical shock and implosion.

The equipment to be used should be chosen as rigorously as that used for work under pressure.

1.1.2.1 Evaporating under Vacuum

Evaporation under vacuum is classically performed using rotary evaporators. Flasks must not be overfilled. Evaporation differs from distillation in that mixtures must not be overheated. Evaporation should be stopped before the flask becomes "dry" in the case where residues may contain unstable products (peroxides, azides, picrates, perchlorates, etc.). The flask should be allowed to cool before the vacuum is broken.

In order to trap solvent vapors which, by passing into the vacuum system, may deteriorate pumps or pollute water (in the case of water aspirators), the efficiency of condensers may be improved by using a cryogenic fluid (e.g., ethylene glycol).

1.1.2.2 Distilling under Vacuum

The apparatus must be adapted to both the quantity and type of product to be distilled. Verification that the apparatus is airtight should be made before each utilization.

The heating element, preferably a heating mantle or a bath (silicon, sand, etc.) should be mounted on a mobile system (lab jack) such that it can be removed quickly if necessary. The use of naked flames for rapid distillations must be accompanied by strict precautions, the systematic use of a safety shield being essential.

Boiling must be regulated by a capillary tube, allowing air or an inert gas (in the case of oxygen or water-sensitive substances) to bubble through. Verification that the capillary has not become obstructed by crystallized product should be made during the course of the distillation. Similarly, the condenser may become obstructed by crystallization of the distilled product.

The distillation flask should not be more than half-filled. It should be placed such that the level of the liquid is below the upper edge of the heating mantle or the bath in order to avoid overheating. Heating must commence only after the vacuum has been applied, otherwise serious bumping and spilling over may occur.

When distillation is complete, allow the contents of the flask to cool before breaking the vacuum with air or nitrogen.

Precautions must be taken when introducing air into a hot flask so that the residue does not catch fire or explode (as in the case of *t*-butyl peroxide,* trimethyl phosphate,† ethyl palmitate, etc.).

The distillation of old compounds likely to have decomposed should be avoided or else performed with extreme caution. A small-scale distillation of these substances should be initially effected before progressively increasing the quantities to be distilled. In these cases, always use a safety shield.

To help avoid accidents or incidents, the literature concerning dangerous chemical reactions should also be consulted beforehand.

In no case should the distillation of mixtures of incompatible substances be attempted (acid chlorides on sodium, etc.).

1.1.2.3 Filtering under Vacuum

Filtration flasks must be in excellent condition and firmly anchored, with undue pressure from clamps being avoided.

If filtration is slow (e.g., colloids), increasing the vacuum will not always improve the situation, though the risk of implosion is increased.

Avoid high vacuums when filtering (at most, 500 mm Hg).

1.1.2.4 Desiccating under Vacuum

Desiccators must be placed in an area where they are unlikely to fall or be bumped. They should be shielded from sunlight, particularly if they contain unstable products. They should never be carried when under vacuum.

Desiccators must be protected by metal screens, thick fabric, adhesive tape, or plastic films of high resistance. Stopcocks and joints should be regularly greased.

A trapping flask must be placed between the desiccator and the water pump in order to avoid explosions due to contact of water with dehydrating agents such as sulfuric acid or phosphoric oxide as a result of back-flow.

The choice of dehydrating agent depends on the product to be dried rather than simply on its dehydrating capacity. Thus, the use of concentrated sulfuric acid should be discouraged owing to its violent reactions with water, bases, amines, and so on. Similarly, phosphoric oxide (P_2O_5), very efficient except in the case of alcohols, amines, and ketones, must be carefully destroyed by adding it in small quantities to water owing to a highly exothermic reaction

---

*Decomposition of t-butyl perbenzoate. L'Actualité Chimique, French Chemical Society, Paris (1983), pp. 49–50.

†Decomposition of trimethyl phosphate. L'Actualité Chimique, French Chemical Society, Paris (1983), pp. 59–60.

which liberates toxic gases. Magnesium perchlorate is not recommended as a desiccant due to its explosive nature.*

1.1.2.5 The Use of Dewar Flasks

Dewar flasks must be protected by either placing them in sturdy containers or covering them with a resistant plastic film. They must not be tied down by chains in direct contact.

Liquids, hot or cold, must be introduced in small portions, care being taken to avoid filling the flask completely or striking the walls. Liquid nitrogen must be introduced in dry Dewar flasks. The progressive absorption of atmospheric oxygen, as indicated by the dark blue color of the mixture, must be avoided by regular replacement of the liquid nitrogen. Beyond an oxygen content of 8%, there is a danger of explosion (see also, cryogenic liquids, p. 29ff).

1.2 Risks Associated with the Physicochemical Properties of Substances

1.2.1 Highly Reactive Substances and Mixtures

1.2.1.1 General Rules

Reactions in which unstable products (or those suspected to be) are involved should only be undertaken with small quantities. Scaling-up must only be done progressively.

Dropwise addition of a solution of the substance in an inert solvent to the reaction mixture is recommended. The reaction must be constantly monitored, particularly with regard to variations in temperature. Excessive cooling can inhibit the reaction and lead to a dangerous buildup of reactants. On the other hand, a sudden elevation in the reaction temperature can lead to loss of control of the reaction. To avoid this latter situation, a cooling mixture should always be kept on hand. Safety shields are recommended.

Certain reactions have varying induction periods, leading to sometimes explosive conditions as with, for example, the oxidation of paraformaldehyde to glyoxal with nitric acid.

Before employing unstable substances and substances of unknown stability or preparing unstable mixtures, a literature search concerning these substances must imperatively be performed.

The principal sources of information regarding chemical reactions and dangerous substances are:

---

*Stability of perchloric acid and perchlorates. L'Actualité Chimique, French Chemical Society, Paris (1986), pp. 41–47.

Bretherick's handbook of reactive chemical hazards, by L. Bretherick. 4th Ed. Butterworths, London (1990).

Hazards in the chemical laboratory, by L. Bretherick. 4th Ed. The Royal Society of Chemistry, London (1987).

Dangerous chemical reactions — A compendium edited by J. Leleu, INRS, Paris (1987) (in French).

Chemical reactions and hazardous substances — Safety section of L'Actualité Chimique, Paris, under the supervision of A. Picot (in French) (1st note September 1983).

Laboratory hazards bulletin (No. 1, July 1981ff). Monthly bulletin published by the Royal Society of Chemistry, Nottingham.

Chemical hazards in chemistry (No. 1, January 1984ff). Monthly bulletin published by the Royal Society of Chemistry, Nottingham.

Occasional explosives, by Louis Medard. Industry – Production – Environment Collection: Techniques and Documentation, Lavoisier, Vol. 1: Properties, Vol. 2: Monograph, 2nd Ed. Paris (1990) (in French).

1.2.1.2 Explosive Substances

Many substances or classes of substances may explode as a result of heating, mechanical shock (striking, grinding), or exposure to light (it is thus dangerous to leave substances of unknown stability in direct sunlight). Such explosions may occur in the presence of catalysts or even spontaneously.

The detonation temperatures of explosive compounds can vary considerably (Table 1.2).

Explosive Compound	Detonation Temperature
Nitroglycerine	177°C
Mercury Fulminate	180°C
Lead azide	350°C
Trinitrotoluene (TNT)	470°C

Table 1.2   Detonation temperatures of several explosive compounds.

Among particularly unstable or explosive families of substances should be noted:

Hydrogen peroxide derivatives.   Hydrogen peroxide, hydroperoxides, peroxides, peracids, peranhydrides, peresters, and so on.

Distillation of liquids belonging to these families of compounds is often delicate.* The formation of peroxides via auto-oxidation of certain solvents

---

*Decomposition of t-butyl perbenzoate. L'Actualité Chimique, French Chemical Society, Paris (1983), pp. 49–50.

(ethers, etc.) or monomers (e.g., 1,1-dichloroethylene and 1,3-butadiene) in air can be a serious danger.

*Perchloric acid derivatives.** Perchloric acid, metallated perchlorates (Ag^+, In^+, Mg^{++}, Ba^{++}, Pb^{++}, Hg^{++}, Co^{++}, Ni^{++}, lanthanides, etc.), perchloryl fluoride, perchlorates of alkanes, amines, hydrazines, pyridines, and so on.

Perchloric acid (70% aqueous solution) can react violently with organic materials (wood, rubber, cork, clothing, etc.). Its vapors are rapidly absorbed by these substances which may become spontaneously flammable or explosive. Perchloric acid should therefore be stored some distance from other substances, preferably in a nonflammable retaining tray. Always store oxidizers, such as perchloric acid and hydrogen peroxide, away from (preferably in a diked area) organics and other reactive materials.

The evaporation of perchloric acid-containing solutions should be performed in a fume-hood specifically designed for perchloric acid use and then only with extreme caution. The efficiency of the solvent extractor and the cleanliness of the apparatus must be regularly checked, otherwise the vapor of this acid may react violently with any residues that may collect in the conduits.

Anhydrous perchloric acid is even more dangerous and explodes either spontaneously or on contact with numerous substances (reducing compounds, combustible materials, etc.).

Magnesium perchlorate must never be used as a dehydrating agent (for alcohols, acetonitrile, DMSO, etc.), for instance, with alcohols, highly explosive alkyl perchlorates may be formed.

Lanthanide perchlorates (erbium, neodymium, etc.) form highly unstable complexes with acetonitrile which can explode simply as a result of scraping with a spatula.

Alkyl perchlorates are particularly hazardous. Ethyl perchlorate is five times more explosive than nitroglycerine.

$$CH_3\text{-}CH_2\text{-}O\text{-}ClO_3$$

Ethyl perchlorate

$$CH_2\text{-}O\text{-}NO_2$$
$$|$$
$$CH\text{-}O\text{-}NO_2$$
$$|$$
$$CH_2\text{-}O\text{-}NO_2$$

Nitroglycerine

Nitrated derivatives. Numerous nitrates, both inorganic (ammonium nitrate) and organic (nitrated esters) are unstable. Highly unstable nitric esters or fulminic acid are easily formed by simple contact between an alcohol and nitric acid. Thus, the use of alcohol to clean glassware which was employed for

---

Stability of perchloric acid and perchlorates. L'Actualité Chimique, French Chemical Society, Paris (1986), pp. 41–47.

nitrations may lead to a violent explosion.

$$R - O - NO_2$$

nitric esters

$$H - C \equiv N^+ - O^-$$

fulminic acid

Numerous polynitrated aliphatic derivatives (trinitroglycerine) or aromatic derivatives (picric acid, picrates, trinitrotoluene, etc.) are very sensitive to mechanical shock and to heat. Heavy metal picrates (Pb, Hg, Cu, Zn, etc.) are particularly unstable. Polynitrated aromatic derivatives can be stabilized by the addition of water (10% or more).

Diazo derivatives. A concentrated solution of diazomethane (CH_2N_2) can explode in the presence of impurities or of traces of solids (the rough surfaces of ground-glass joints sometimes suffice). Diazomethane must be dried with solid potassium hydroxide rather than with calcium sulfate.

Diazonium salts are very often unstable, especially when dry (dry or zinc chloride-complexed diazonium chlorobenzene and diazonium perchlorates).

Hydrazoic acid derivatives. A large number of inorganic (Ag, Hg, Pb, Cu, etc.) and organic azides are very unstable. Highly explosive aryl azides may form in diazotization reactions such as the Sandmeyer reaction.

Highly explosive silver azide (AgN_3) is formed by prolonged storage of solutions of silver nitrate and ammonia (Tollens reagent). This reagent must thus be prepared just before use and the correct concentrations respected. It may be destroyed by the addition of a solution of aqueous sodium chloride.

Sodium azide, used as a bactericidal agent for aqueous solutions stored in metallic (particularly copper) containers, can form explosive metallic azides.

Acetylenic derivatives. Acetylene, dichloroacetylene, polyacetylenic derivatives, acetylenic derivatives, acetylides.

Acetylene and heavy metal acetylides (Ag, Cu) may explode either spontaneously or upon mechanical shock.

Table 1.3 summarizes the principal unstable functional groups that are commonly encountered in the laboratory.

1.2.1.3 Unstable Substances

Prolonged storage of certain unstable products can result in their decomposition, leading to mixtures that may explode under certain circumstances (mechanical shock, heating, or simple handling). This is true of alkaline amides ($NaNH_2$) and certain diazonium salts. Also, aluminum chloride ($AlCl_3$) slowly accumulates hydrochloric acid as a result of its decomposition in the presence

TYPE OF BOND	FUNCTIONAL GROUP	EXAMPLES
$-O-O-$	Peroxy	Hydrogen peroxide
$^-O-O^-$	Metallated peroxide	Sodium peroxide
$-O-O-H$	Hydroperoxide	t-butyl hydroperoxide
$R-O-O-R'$	Peroxide	Di-t-butyl peroxide
$R-C-O-O-R'$ (with $\parallel$ O below C)	Peracid / Peranhydride / Perester	Peracetic acid / Peracetic anhydride / t-butyl perbenzoate
$R-O-X$	Hypohalite	t-butyl hypochlorite
$-O-O-O-$	Ozonide	Ozone
ClO_4^-	Perchlorate	Perchloric acid / Inorganic perchlorates / Organic perchlorates
$R-O-N^+$ (with $=O$ and O^-)	Organic nitrate	Trinitroglycerine
$(R)\left(-N^+{<}^O_{O^-}\right)_n$	Polynitro	Tetranitromethane / Trinitrotoluene
$O^--N^+\equiv C^-\ M^+$	Fulminate	Mercury fulminate
$>N-X$	Haloamine / Haloimide	Chloramine / N-Bromoacetamide
$-N^+\equiv N\ X^-$	Diazonium salt	Benzenediazonium chloride
$N^-=N^+=N^-\ M^+$	Azide	Silver azide / Iodine azide
$-C\equiv C^-\ M^+$	Acetylide	Silver acetylide

Table 1.3 Principle unstable functional groups.

of moisture absorbed during its preparation. An attempt to open the flask may lead to its shattering due to the pressure of the gases which has developed.

The opening of flasks with ground-glass joints whose stoppers are frozen must always be done with caution. Quite frequently, these flasks are under pressure due to the substance contained or to decomposition products formed during storage.

Unstable liquids may be stored in sealed tubes. This is the case for borohalides (BCl_3, BBr_3), for halides and oxyhalides of phosphorus (PCl_3, $POCl_3$), sulfur (SCl_2, $SOCl_2$), tin ($SnCl_4$), antimony ($SbCl_5$), and so on.

Rules for opening ground-glass flasks and sealed ampules.*

−Wear protective goggles or, better, a face screen.
−Work under a fume-hood and behind a polycarbonate safety shield.
−Wear thick gloves.
−Open the flask inside an unbreakable container made of material compatible with the contents of the flask. Wrap a rag around the flask.

If the liquid contained has a boiling point lower than room temperature, it is necessary to cool the sealed ampule in a dry ice bath before opening.

1.2.1.4 Polymerizable Substances

Certain monomers can sometimes spontaneously polymerize explosively or with shattering of flasks or bottles. Such monomers include vinyl acetate, acrolein, methyl acrylate, acrylonitrile, 1,3-butadiene, vinyl chloride, vinylidene chloride, ethylene oxide, β-propiolactone, styrene and so on. Some monomers such as acrylamide polymerize violently upon heating, however, more often polymerization is initiated by trace impurities such as:

−acids (hydrogen cyanide, aziridine, etc.)
−bases (cyanamide, acrolein, etc.)
−metals (acetaldehyde, etc.)

Physical causes such as mechanical shock or exposure to light may also provoke violent polymerizations.

Monomers, preferably stabilized, should be stored in small quantities. Storage close to substances likely to release traces of acids or bases should be avoided.

1.2.1.5 Hazardous Mixtures

Oxidizing agents

The majority of accident-producing chemical reactions involve oxidation-reduction phenomena in which strong oxidants intervene. Always store oxidizers in a separate diked area. Oxidants do not necessarily contain oxygen and their oxidizing power varies with their structure and the reaction conditions (see Table 1.4).

Certain oxidizing agents are particularly dangerous.

Fluorine. Fluorine is the most electronegative and most reactive element. It often reacts violently with most compounds. Thus, bromine and iodine burn at room temperature in the presence of fluorine. It reacts explosively in the presence of water and various solvents (CCl_4, $CHCl_3$, hexane, etc.).

Ozone. In the solid or liquid state, ozone is highly explosive. It reacts with numerous unsaturated compounds (alkenes, arenes) by formation of explosive ozonides.

---

*Reactions in sealed ampules. L'Actualité Chimique. French Chemical Society. Paris (1985). pp. 71–72.

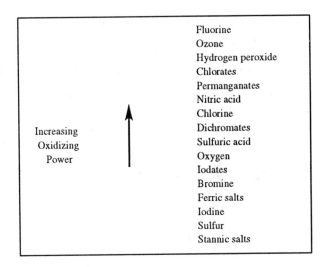

Increasing Oxidizing Power ↑	Fluorine
	Ozone
	Hydrogen peroxide
	Chlorates
	Permanganates
	Nitric acid
	Chlorine
	Dichromates
	Sulfuric acid
	Oxygen
	Iodates
	Bromine
	Ferric salts
	Iodine
	Sulfur
	Stannic salts

Table 1.4   Relative oxidizing strength of various inorganic compounds.

Hydrogen peroxide.   Commercial 30% solutions of hydrogen peroxide are relatively stable and only slowly decompose (but should nevertheless be stored in a refrigerator). Alternatively, more concentrated solutions, in particular those of 90%, are extremely dangerous (these concentrations of hydrogen peroxide react violently in the presence of organics). Trace metals (Fe, Cu, Cr, Pb) decompose hydrogen peroxide with explosion.

Certain derivatives of hydrogen peroxide are powerful oxidants (peracids) and many, such as hydroperoxides, peroxides, and peresters, are unstable.

Alkaline chlorates.   Sodium and potassium chlorates ($NaClO_3$, $KClO_3$) are powerful oxidants. Many compounds (carbon, sulfur) burn spontaneously in the presence of chlorates. Chlorates must not be dried on paper; spontaneous combustion due to static electricity is possible.

Potassium permanganate.   Potassium permanganate ($KMnO_4$) is a powerful oxidizing agent in acid, basic or neutral media. In the solid state, it can cause certain organic compounds such as ethylene glycol, glycerol, acetic acid, acetic anhydride, and so on, to burn. The addition of concentrated sulfuric acid to a concentrated solution of potassium permanganate can lead to explosion because of the formation of permanganic acid ($HMnO_4$).

Nitric acid.   Nitric acid (HNO_3) is both a strong acid and a strong oxidizing agent. Its oxidizing power and its instability increase with concentration (concentrated acid: 70%, fuming acid: 85%, anhydrous acid: 100%).

Nitric acid forms unstable nitrates with a large number of compounds such as alcohols, acetic anhydride, acetone, acetonitrile, and so on. The use of nitric acid or sulfonitric acid (HNO_3 + concentrated H_2SO_4) for the cleaning of glassware is not recommended because of the danger of explosion.

Chlorine.   Chlorine (Cl_2) is a very aggressive gas that explodes when mixed with hydrogen or hydrocarbons in the presence of light. It can also lead to spontaneous combustion of ethyl ether and, moreover, reacts violently with numerous compounds (diborane, *t*-butanol, hydrocarbons, synthetic rubbers, etc.).

Dichromates.   Alkaline dichromates are powerful oxidizing agents in acidic media. Mixed with sulfuric acid, they are still used for the cleaning of glassware. The sulfochromic mixture is prepared by dissolving either potassium dichromate ($K_2Cr_2O_7$) or sodium dichromate ($Na_2Cr_2O_7$) in concentrated sulfuric acid.

These dichromates, like all hexavalent chromium salts, are powerful irritants, allergens, mutagens, and carcinogens, both in laboratory animals and in humans. They are responsible for a certain number of professional illnesses. They are also major pollutants of the aquatic environment and should, therefore, never be washed down a sink.

The sulfochromic mixture should, as often as possible, be replaced by other types of cleaning solutions. Depending on the material to be treated, the least dangerous replacement should always be chosen.

Among the replacement solutions the following may be considered:

–concentrated alkaline solutions of surface-active agents such as TFD_4 or RBS. These very basic solutions (pH 13) must be handled cautiously and protective goggles should always be used;

–ultrasonic baths, keeping in mind the safety measures applicable to the material to be introduced;

–ammonium peroxodisulfate (ammonium persulfate) in sulfuric acid may be used in exceptional cases and as diluted as possible, the instability of this mixture requiring strict precautions. This mixture may be used at different concentrations.

Chromic acid.   Chromic acid (or chromium trioxide, CrO_3) reacts with a large number of organic compounds (acetic acid, acetic anhydride, acetone, alcohols, pyridine, DMF, HMPT, etc.). A solvent must never be added to solid chromic acid due to the serious risk of explosion and fire.

Ammonium persulfate.   Ammonium persulfate or ammonium peroxodisulfate [$(NH_4)_2S_4O_8$], a derivative of hydrogen peroxide, is an unstable compound that explodes when exposed to heat, friction, or mechanical shock, liberating highly corrosive gases. According to H.M. Stahr et al,* a mixture suitable for cleaning glassware (from which most of the solid residues have been removed) can be prepared by cautiously pouring 5 g of ammonium persulfate into a Pyrex container of at least 2 l containing 500 ml of concentrated sulfuric acid. The container should be kept in an ice bath and the mixture stirred until all the solid has dissolved. The solution can then be left to come to room temperature before use.

*H. M. Stahr. *Anal. Chem.* 1982, **54**, 1456A.

This cleaning mixture should be periodically reactivated by addition of 100 ml of H_2SO_4 containing 0.8 g of ammonium persulfate. Another method of preparation has also been proposed by Stahr and collaborators (previous reference). In a Pyrex flask of at least 5 l, 95 g of ammonium persulfate is introduced to which is slowly added, at room temperature, 2 l of water. The mixture is stirred until all the solid has dissolved. The solution is then cooled in an ice bath before carefully adding 1.25 l of concentrated sulfuric acid. The solution is allowed to come to room temperature.

All these preparations and operations must be effected under an efficient fume-hood and using a safety shield, a face shield or, at the very least, protective goggles. Acid-proof gloves and, if possible, an acid-proof apron or a lab coat should be worn.

The least powerful, and thus most stable, mixture should first be tested.

Oxygen.   Oxygen (O_2) is a powerful oxidizing agent in the presence of catalysts. Serious explosions can occur as a result of contact of oxygen gas under pressure with organic substances.

Oxygen cylinder valves must never be greased.

Numerous organic compounds such as ether, tetralin, decalin, isopropanol, autoxidize in air, particularly in the presence of light, leading to the formation of explosive peroxides.

Reducing agents

Strong reducing agents such as hydrogen, hydrazine, and mixed hydrides, can explode or burn, especially in the presence of impurities.

Hydrazine.   Hydrazine (NH_2NH_2), a powerful reducing agent, is particularly dangerous. Hydrazine, as well as its hydrate (64% hydrazine), decompose readily when heated. The flash points in air are 270°C for hydrazine and 292°C for hydrazine hydrate. Numerous porous substances, for example, asbestos or metal powders, accelerate this decomposition, leading to explosion. Hydrazine vapors can spontaneously ignite in air.

The preparation of anhydrous hydrazine, whose lower and upper flammability limits (lfl and ufl or lel and uel) in air are 4.7 and 100%, by distillation of the hydrate in the presence of potassium hydroxide (CaO and BaO are to be avoided) must be performed under a nitrogen or argon atmosphere. Azeotropic distillation — with aniline, for example — is less dangerous. Hydrazine reacts violently with most oxidizing agents such as chromium salts (chromates, chromic acid), potassium permanganate, hydrogen peroxide, and nitric acid. Hydrazine also reacts with numerous metals (iron, platinum, mercury) and their oxides.

Lithium aluminum hydride.   Lithium aluminum hydride ($LiAlH_4$), a mixed hydride, is used as a powerful reducing agent. In the solid state, it may spontaneously burn in air, particularly in humid air.

The addition of powdered lithium aluminum hydride to a distillation flask containing an ether to be purified (such as diethyl ether, 1,2-dimethoxyethane, THF, or dioxane) can cause spontaneous ignition of the mixture. In accidents of this type, the fire must be extinguished with powder extinguishers and never with water or CO_2 extinguishers.

The reduction of numerous compounds with $LiAlH_4$ can lead to explosive reactions as, for example, with benzoyl peroxide and benzaldehyde. Its destruction by progressive addition of ethyl acetate must be effected with caution. $LiAlH_4$ must be stored in small quantities in a desiccator, protected from air and moisture.

Substances reacting with water

Numerous substances react violently and sometimes explosively with water. The use of small quantities of reactive substances or of water favors violent reactions. Several classes of substances react with water.

Alkali metals.   Cesium, rubidium, and potassium explode violently in the presence of water. The hydrogen formed ignites spontaneously. Sodium reacts less violently and the liberated hydrogen does not ignite. In the presence of ice, however, sodium will explode.

The reaction of lithium with cold water is moderate, but becomes violent with hot water, and the liberated hydrogen can ignite.

Calcium.   Calcium also reacts violently with water.

Alkali hydrides and alkaline earth metals.   The hydrides of sodium (NaH), potassium (KH), lithium (LiH), and calcium (CaH_2) react violently with water, liberating hydrogen which spontaneously ignites.

Certain mixed hydrides.   Lithium aluminum hydride ($LiAlH_4$) reacts very violently with water and must not come in contact with moist solvents.

Calcium carbide.   Calcium acetylide or calcium carbide (CaC_2) reacts with water, liberating acetylene which can ignite, especially if under pressure.

Alkaline amides.   Potassium amide (KNH_2) and sodium amide ($NaNH_2$) react with water. Old, degraded samples, which can be recognized by their yellow color, do not react immediately with water, but a violently explosive reaction occurs soon after.

Organometallics.   Certain organometallic compounds, alkali metals such as methyllithium and n-butyllithium, alkaline earth metals (organomagnesiums), compounds of aluminum, zinc, and cadmium (dimethylcadmium) react violently with water. Their destruction in water must be effected cautiously by slow addition of the organometallic compound to ice water.

Diborane.   Diborane (B_2H_6) reacts vigorously with water and may burn. It also decomposes at ambient temperature, liberating hydrogen and can ignite when in contact with air. It must not be diluted with carbon dioxide with which it can react. Borohydrides react more slowly, except for lithium borohydride ($LiBH_4$).

Halides derived from nonmetallic elements

Boron halides.   Boron halides, such as boron trichloride (BCl_3), boron triiodide (BI_3), and boron tribromide (BBr_3) react vigorously with water. This reaction is particularly violent (even in the case of small quantities of water) with boron tribromide.

Phosphorus halides and oxyhalides.   Phosphorus trichloride (PCl_3) reacts violently with water, liberating diphosphane which ignites spontaneously. Phosphorus pentachloride (PCl_5) also reacts violently with water and the products of hydrolysis can, in turn, react violently with an excess of water. Phosphorus oxychloride ($POCl_3$) exhibits delayed but often violent reaction with water.

Sulfur halides and oxyhalides.   Sulfur dichloride (SCl_2) reacts with water, producing heat. Thionyl chloride ($SOCl_2$) and sulfuryl chloride (SO_2Cl_2) react violently with water, liberating much gas ($SO_2 + HCl$) in the case of thionyl chloride. Similarly, chlorosulfonic acid (HSO_3Cl) reacts very violently with water, the reaction being, in addition, very exothermic.

Phosphoric oxide.
Commonly used as a desiccating agent, phosphoric oxide (phosphorus pentaoxide, P_2O_5) reacts energetically with water, liberating gases which can ignite, the reaction being exothermic. When used as a desiccating agent, phosphoric oxide becomes covered with a viscous, protective film of phosphoric acid which retards its destruction with excess water, thereby increasing the risk of a delayed, violent reaction.

Calcium oxide.
Calcium oxide (or lime, CaO) reacts with water, liberating considerable heat. Storage in a glass flask in a humid environment can lead to explosion of the flask due to the exothermic hydration of calcium oxide in the presence of traces of water.

Sodium peroxide.
The violence of the reaction of sodium peroxide (Na_2O_2) with water is directly related to the quantity of the latter.

Potassium superoxide.
Potassium superoxide (KO_2) also reacts violently with water.

Carboxylic acid halides and anhydrides.
Acetyl chloride, acetyl bromide, and acetic anhydride react violently with water, liberating the corresponding carboxylic acid.

General recommendations for the handling of substances reacting with water:

- Work under a fume-hood with a safety shield, eliminating all unnecessary flammable materials from the work area.
- Wear protective goggles with sideshields at all times.

- Manipulate alkali metals with forceps and, for potassium, work under Vaseline oil or xylene in a dry mortar. Under prolonged contact with water, potassium forms potassium superoxide which may explode when fresh potassium is exposed as when it is cut with a scalpel.
- Work with small quantities.
- Always introduce the reactive compound into water and never the reverse. Thus, the dilution of concentrated mineral acids (H_2SO_4, oleum, HNO_3, HCl, HBr) must always be effected by addition of the acid to small quantities of cold water. Similarly, the addition of strong bases to water is very exothermic (the heat of solution of sodium hydroxide is of the order of 10 kcal/mole).
- Residues of alkali metals and their derivatives should be destroyed before they accumulate by adding them to a higher alcohol (*n*-butanol) in an appropriate vessel such as a Petri dish.
- After use, reactive products must be stored in cupboards reserved for this purpose.
- Certain highly reactive compounds (hydrides, amides, $LiAlH_4$) must be stored in desiccators.

The quantities of highly reactive substances stored should be minimized.

Remark: Certain reactive compounds such as amides do not withstand prolonged storage (e.g., 2 months for $NaNH_2$).

Regularly verify the state of stored reactive chemical products.

Pyrophoric substances

Many types of compounds can spontaneously ignite in air and are referred to as pyrophoric:

- Finely powdered metals such as Raney nickel, reduced platinum, palladium black, zinc, and so on.
- Simple hydrides such as diborane (B_2H_6) and sodium hydride (NaH) or mixed hydrides such as lithium borohydride ($LiBH_4$), and so on.
- Hydrogen phosphide or phosphine (PH_3) and substituted phosphines such as trimethylphosphine [$P(CH_3)_3$].
- Low molecular weight silanes: SiH_4, Si_2H_6.
- Organometallics such as trimethylaluminum [$Al(CH_3)_3$], trimethylbismuth [$Bi(CH_3)_3$] and heteroatomic organic compounds such as trimethylboranes [$B(CH_3)_3$] and trimethylarsine [$As(CH_3)_3$].
- Metal carbonyls such as iron carbonyl [$Fe(CO)_5$] and nickel carbonyl [$Ni(CO)_4$].
- White phosphorus liberates phosphine (PH_3), among other substances, when added to a hot solution of strong base. The liberated phosphine also ignites spontaneously in air.

All pyrophoric substances must be handled under an inert atmosphere (dry nitrogen, argon, etc.).

The residues of pyrophoric substances must never be disposed of in a trash can but must be destroyed using appropriate techniques.

Other hazardous mixtures

The mixing of certain compounds or the preparation of unstable products may lead to dangerous reactions. As examples may be cited:

The preparation of nitrogen trihalides. Ammonia (NH_3) reacts with halogens (Br_2, Cl_2, and I_2) or their derivatives (NaOCl) with formation of highly unstable nitrogen trihalides (NX_3). Thus, a mixture of ammonia and iodine forms nitrogen triiodide (NI_3) which explodes under the least handling. All nitrogen trihalides (NX_3) are unstable in the presence of numerous compounds, including oil present on fingers.

Addition of anhydrides and acid halides to dimethyl sulfoxide. In the absence of a diluting agent, dimethyl sulfoxide (DMSO), a dipolar, aprotic solvent, reacts violently with a variety of acid anhydrides or acid halides such as:

- acetyl chloride
- trifluoroacetic anhydride
- benzenesulfonyl chloride
- PCl_3, $POCl_3$
- SCl_2, S_2Cl_2, SO_2Cl_2, $SOCl_2$

Table 1.5 lists the principal incompatible mixtures.

1.2.2 Cryogenic Fluids

The use of cryogenic fluids (air, nitrogen, oxygen, argon, helium, neon, hydrogen) necessitates appropriate equipment, resistant both to the action of the gas and to low temperatures. The transfer of cryogenic fluids should be effected by siphoning whenever operating conditions permit. For cryogenic fluids that evaporate constantly (air, nitrogen, oxygen, etc.), a venting orifice should be present and periodically verified for obstruction.

Their use entails the following risks:

Fire. Certain cryogenic liquids, such as hydrogen, methane, and acetylene, are extremely flammable.

Explosion. Oxidizing cryogenic fluids such as oxygen and liquid air can lead to violent reactions in the presence of reducing agents. Liquid nitrogen or argon, when left in contact with air, become oxygen enriched which may be dangerous upon warming.

Body lesions. Contact with skin or mucosa, particularly eyes, can provoke serious burns comparable to thermal burns.

Products[1]	Incompatible products[1]	Exothermic reaction[2]	Explosive reaction	Spontaneous ignition	Toxic gas formation
Acetylene Acetylenic derivatives	Silver, Mercury, Copper		+		
Strong mineral acids	Water Mineral bases Cyanides Azides Sulfides Hypochlorites	+ +			+ + + +
Strong mineral bases	Water Strong acids Phosphorus	+ +			+
Bromine Chlorine	Unsaturated compounds Carbonyl compounds Diethyl ether Ammonia Phosphane Silane Phosphorus	+ + + +	+	+ + + +	
Organo-metallic compounds	Water Air Oxygen	+ + +		+	
Non-metallic hydrides SiH_4, PH_3	Air Oxygen	+ +		+ +	
Alkaline hydrides and alkaline earth hydrides	Air Oxygen Water	+ + +		+ + +	
Mercury	Acetylene Ammonia Halogens Alkaline metals Sulfur	+ + + +	+		

Table 1.5 Principle groups of incompatible substances.

Products[1]	Incompatible products[1]	Exothermic reaction[2]	Explosive reaction	Spontaneous ignition	Toxic gas formation
Alkaline metals	Water	+		(+)	
	Alcohols	+		(+)	
	Halogens	+	+		
	Halides	+			
Metal carbonyls	Air			+	+
	Oxygen			+	+
Nitroalkanes Nitroarenes	Mineral bases		+		
Powerful oxidizing agents $KMnO_4$, O_3, Cr^{VI} salts, H_2O_2)	Unsaturated organic compounds	+		(+)	
	Reducing agents	+	(+)	(+)	
Phosphorus	Air			+	
	Oxygen			+	
	Mineral bases			+	+
	Oxidizing agents		+		
	Halogens	+		+	

(1) or chemical families   *(2) or formation of an explosive product*

Table 1.5 (continued)   Principle groups of incompatible substances.

Intoxication.   Most cryogenic fluids are nontoxic or only slightly toxic (see Table 1.6). However, the evaporation of nitrogen, for example, in confined quarters may lead to asphyxiation and anoxia due to the resulting displacement of oxygen. The same considerations apply to carbon dioxide (CO_2). Inversely, breathing an oxygen-enriched mixture can provoke physiological problems in direct proportion to the partial pressure of the oxygen inhaled.

Cryogenic fluids commonly used in the laboratory include the following:

1.2.2.1 Solvents Cooled by Dry Ice

The choice of solvent for the preparation of a dry ice cooling bath depends on the temperature required. Care must be taken, however, that the solvent employed cannot react with the reaction mixture should the immersed vessel break. The use of organic solvents such as acetone, for example, should be avoided when reactions with oxidizers such as hydrogen peroxide are performed since the formation of acetone-derived peroxides is possible as well as violent reactions if the hydrogen peroxide concentration is significant. Simi-

Gas	Boiling Points	Volume of Expansion	Toxicity
Acetylene	- 84	-	+
Hydrochloric acid	- 85	-	+
Nitrogen	- 195	696 to 1	-
Argon	- 185	847 to 1	-
Carbon dioxide	- 78	553 to 1	+
Helium 3	- 269	757 to 1	-
Helium 4	- 268	757 to 1	-
Hydrogen	- 252	851 to 1	-
Methane	- 161	578 to 1	-
Carbon monoxide	- 192		+
Oxygen	- 183	860 to 1	-
Boron trifluoride	- 100		+

Table 1.6   Properties of cryogenic fluids.

larly, acetone reacts violently with sulfonitric mixtures. The addition of dry ice to solvents must be done with caution and in small quantities in order to avoid all possibility of spills.

1.2.2.2 Liquid Nitrogen

Particular attention must be paid to the dangers associated with oxygen-enrichment of liquid nitrogen traps, the condensation of redox mixtures increasing the risk of explosion (see Section 1.1.2.5).

1.2.2.3 Liquid Oxygen

Liquid oxygen forms explosive mixtures with numerous solid materials, particularly porous ones. Its manipulation must be accompanied by stringent precautions.

1.2.2.4 Recommendations for the Use of Cryogenic Fluids

–Constantly wear safety goggles and, if possible, work behind a screen.
–Check the quality of the equipment, especially of the glassware in contact with the cryogenic fluid.
–Keep the vessel to be cooled (Dewar flask) in a stable, solid protective envelope.
–Wear temperature-resistant gloves.
–For large quantities, verify that the work area is well ventilated.

1.2.3 Flammable Substances

The risks associated with the flammability of a substance or a mixture can be evaluated by means of the relative importance of different phenomena or situations.

1.2.3.1 Self-ignition Temperatures

This represents the minimum temperature at which a combustible mixture, at a given pressure and composition, spontaneously ignites in the absence of a flame. This phenomenon of self-ignition occurs with a delay ranging from a few milliseconds (at elevated pressure and temperature) to several minutes (at low pressure and temperature), for example, carbon disulfide ignites spontaneously when heated to 120°C.

1.2.3.2 Flash Point

This is the lowest temperature at which the concentration of emitted vapors is sufficient to produce deflagration when exposed to a flame or a hot spot but insufficient to assure propagation of combustion in the absence of the "pilot" flame. The presence of static electricity is particularly dangerous.

1.2.3.3 Flammability (or Explosion) Limits (UFL and LFL or UEL and LEL)

Combustion can only be maintained and propagated if the concentration of combustible material in a gaseous mixture (usually air) is situated between two values:

The *lower flammability (or explosion) limit (UFL or UEL)* of a mixture corresponds to the concentration of combustible material below which combustion can neither be maintained nor propagated. Fire depends on a triangle of three key ingredients: (1) oxygen, (2) fuel, and (3) heat.

Below the LFL there is insufficient fuel to feed the fire or explosion. For a particular combustible material, this value can vary greatly depending on:

- the pressure of the mixture
- the initial temperature
- the concentrations of inert gas in the mixture.

The *upper flammability (or explosion) limit (UFL or UEL)* of a mixture corresponds to the concentration of combustible material above which combustion can neither be maintained nor propagated. Above the UFL there is insufficient oxygen to feed the fire.

These flammability (or explosion) limits are expressed as a percentage of the product vapors in air. For example, the explosion limits for several common flammable substances are given in Table 1.7.

COMPOUND	EXPLOSION LIMITS	
	LOWER (%)	UPPER (%)
Acetone	2.6	13
Acetylene	2.5	81
Benzene	1.4	8
Cyclohexane	1.3	8.3
Dioxane	1.7	22
Ethanol	3.3	19
Diethyl ether	1.9	36
Diisopropyl ether	1.4	8
n-Hexane	1.2	7.5
Hydrogen	4.0	75
Methanol	4.0	36
Methyl ethyl ketone	1.8	11.5
Carbon disulfide	1.3	50
Tetrahydrofuran	2.0	11.5
Toluene	1.3	7

Table 1.7   Explosiveness limits of several chemical compounds.

1.2.3.4 *The Rate of Evaporation*

This refers to the rate of evaporation of a substance compared to that of ether:

$$\text{Volatility} = \frac{\text{Rate of evaporation of substance}}{\text{Rate of evaporation of ether}}$$

The *maximum vapor pressure* may also be taken into account, this being the pressure exerted on a liquid by its emitted vapors once dynamic equilibrium between the liquid and gas phases has been attained.

Volatility and maximum vapor pressure should be considered when estimating not only the risk of ignition but also the danger of product toxicity since they allow the determination, under given conditions, of the quantity of emitted vapors as well as of their rate of emission.

Remark: The storage of volatile and flammable liquids in nonexplosion-proof refrigerators should be avoided.

All refrigerators must be equipped with an exterior thermostat and have no interior light fixtures.

Flammable products may be solid (powdered metals), liquid (most solvents), or gas (the most hazardous).

–Never manipulate flammable products near a flame or hot spot.
–Avoid storing flammable solvent containers on shelves above the work-

bench. These should be stored in a ventilated cupboard or cabinet away from evacuation routes.

Never store large quantities of flammable substances in the laboratory. Check with local fire code regulations (typically based on NFPA 30) to ensure the laboratory is not exceeding quantity limitations.

1.3 Risks Associated with the Toxic Properties of Substances

Toxic effects are the result of an interaction between a xenobiotic substance and a living organism (or ecosystem) (Figure 1.1).

The toxicity of a given substance depends on:

- The quantity introduced into the organism (dose). (The dose makes the poison. Depending on the dose almost any compound can be toxic, even water.)
- The cumulative character of the doses or of their effects
- The penetration route
- The metabolic capacities of the individual
- The state of health of the subject
- The state of the subject (fatigue, stress, etc.) at the time of exposure
- The presence of other substances introduced accidentally or otherwise into the organism (other toxic substances, tobacco, alcohol, narcotics, medication, etc.).

Overall, the toxicity of a given substance from the moment of its introduction into an organism (exposure phase) to the time of an observable effect

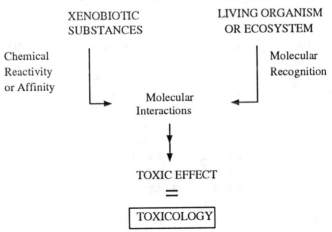

Figure 1.1   The expression of toxicity.

(toxicodynamic phase) may be divided into three major steps in which metabolism (toxicokinetic phase) plays an essential role:

- an exposure phase during which the xenobiotic substance is present in an absorbable form (gas, aerosol, vapor, liquid, solid, etc.);
- a toxicokinetic phase which corresponds to the absorption of a substance by a variety of routes followed by its redistribution in the organism, mainly via blood. If the substance is not directly rejected unchanged or stored in some compartment of the organism, it may be subjected to enzymatic conversions, transforming it into one or several water-soluble metabolites before ultimate excretion;
- a toxicodynamic phase during which the xenobiotic itself (direct toxin) or the "ultimate" toxin resulting either from hydrolysis or metabolism of the xenobiotic (metabolite) interacts with its biological targets.

1.3.1 Routes of Entry

Toxic substances may enter an organism by four main routes: pulmonary, oral, transcutaneous, and percutaneous to which may be added the particular case of ocular contact.

A general rule correlating the seriousness of an intoxication with the route of toxin penetration cannot be formulated. Each toxic substance is more or less dangerous depending not only on its mode of penetration but also on its solubility, its physical state, and, of course, its intrinsic toxicity.

1.3.1.1 Inhalation: Pulmonary Route

The respiratory system allows the filtration of several thousand liters of air per day (2,000 l per 8 hour day for a sedentary worker and up to 1,000–1,500 l per hour for a manual worker) and this over a very large exchange area (the alveolar surface area is $90\,m^2$ at rest and $130\,m^2$ during deep inhalation) making it the principal, and often the most sensitive, penetration route for toxic substances in the organism.

Thus, upon intoxication, while the degree of exposure is easily determined, the actual quantity of substance inhaled is a function of respiratory parameters, of the physicochemical properties of the substance (see Section 1.2.3.4 for explanations concerning the rate of evaporation and vapor pressure), and, for aerosols and particles, of its dimensions. Dust particles larger than 5μ are trapped by the nose, trachea, bronchial tubes, and mucosa and expelled by means of hairs which coat these passageways. Tobacco smoke which perturbs the functioning of these hairs is an aggravating factor in the case of intoxications due to dust.

It should be noted that certain toxic substances diffuse directly to their target organs without passing through the liver. This is the case for many gases and solvents, whose acute toxicities are consequently increased.

It is the air concentration of a toxic substance which is used to determine toxic threshold levels. Most tables present maximum permissible concentrations for exposures of 40 hours per week. These threshold concentrations presumably result in no observable clinical symptoms. They are regularly revised when new toxicological data or new techniques of detection or protection become available. During the past few years, toxic threshold levels for single exposures have also appeared (Immediately Dangerous to Life and Health or IDLH). The units employed are typically expressed as ppm (parts per million) or mg/m^3.

$$\text{Value in ppm} \times \frac{\text{Molecular weight}}{24.45} = \text{Value in } mg/m^3$$

(at 25°C and 760 mm Hg).

The American Conference of Governmental Industrial Hygienist's Threshold Limit Values (TLVs) are not legal requirements. Occupational Safety and Health Authority's (OSHA's) Permissible Exposure Limits (PELs) are legal requirements. Both TLVs and PELs pertain to the inhalation of a single toxic substance during a given period of exposure. However, formulas permit the evaluation of mixtures or the successive inhalation of several toxic substances.

Circumspection is required when considering the possibilities of olfactory detection of toxic substances. These vary greatly from individual to individual as well as for the same individual, depending on his immediate dispositions. Moreover, many compounds are able to anesthetize olfactory cells, incapacitating this mode of detection. The efficiency of the olfactory system also varies considerably as a function of the family of compounds to be detected. Many compounds have odor thresholds higher than their PEL, allowing individuals to unknowingly be exposed to concentrations above the safe threshold.

There is no relationship between toxicity and odor.

1.3.1.2 Transcutaneous Route

This is the second most common mode of penetration of toxic substances associated with professional activities. Cutaneous absorption is a function of the integrity of teguments and, more importantly, of the physicochemical properties of the substance. Effects can be local (allergy, necrosis) or generalized. Certain compounds act as sensitizers, others are absorbed very rapidly and may even serve as vehicles (as with dimethyl sulfoxide, DMSO), allowing substances to penetrate skin which alone would not normally do so.

All organic solvents able to dissolve grease may be suspected of transcutaneous toxicity.

The wearing of appropriate gloves with the best breakthrough times (i.e., the longest) should be required and the use of creams (barrier creams) must be prohibited.

1.3.1.3 Percutaneous Route

Wounds caused by glassware (particular care should be given to certain operations such as the introduction of glass tubes into flexible tubes) or needle pricks are the principal sources of intoxication by this route. Infections should be rapidly treated. While, this type of intoxication may be rare except with highly toxic substances (acutely toxic and biologically active materials), the possibility of chemical exposure should not be ignored.

1.3.1.4 Oral Route

This route of entry is professionally infrequent. Except in certain rare cases, the toxin passes through the liver which is, as will be seen, the principal organ of inactivation and transformation (though at the same time, the principal target organ). Typically, this route of exposure occurs from poor hygiene (i.e., allowing employees to eat while working, not emphasizing that employees should wash their hands before eating, not enforcing a glove policy, etc.).

> A strict application of the rules against pipetting or siphoning all substances by mouth and of eating, drinking, or smoking in laboratories is sufficient to eliminate all risks of oral penetration.

It should be noted, however, that certain inhaled substances may be naturally rejected toward the back of the throat, but in doing so, lead to secondary oral intoxication.

1.3.2 Types of Toxicity

The different types of toxicity may be classed either as:

- a function of the mode of intoxication associated with the reactivity of the chemical compound. This is referred to as direct or indirect toxicity, or
- a function of the delay in the appearance of clinical symptoms, of their nature and/or of the quantities absorbed. Toxicity may then be classified as acute or subacute which may, in turn, be more or less long term (Figure 1.2).

Long-term intoxication is often incorrectly referred to as "chronic intoxication." It should be noted that there is generally no relation between acute toxicity and long-term toxicity.

1.3.2.1 The Classification of Toxicity as a Function of Reactivity

Direct toxic agents

Substances demonstrating high chemical reactivity act directly on the organism

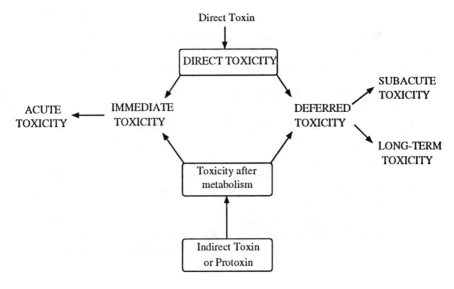

Figure 1.2   Principal forms of toxicity.

without need of any type of transformation. These toxic substances act directly and generally rapidly on the target organ. This is the case of numerous toxins (botulism toxin, etc.), alkaloids (strychnine, brucine, nicotine, etc.), highly reactive alkylating agents (dimethyl sulfate, methyl fluorosulfonate, diazomethane, ethylene oxide, aziridine, formaldehyde, etc.). The majority of corrosive products (acids, bases, strong oxidizing agents, etc.) or irritants (quinones, etc.) are direct toxic agents.

Remark: Water-sensitive compounds that hydrolyze on contact with water in the organism, liberating one or more highly reactive substances, may be classified as direct toxic agents. This applies to certain acid halides (phosgene, see Fig. 1.3, halides of boron, phosphorus, and sulfur), nitrosoureas, nitrosoguanidines.

$$\underset{\text{Phosgene}}{\overset{\displaystyle Cl}{\underset{\displaystyle Cl}{>}}C = O} \;+ H_2O \longrightarrow \left[\underset{\text{Carbonic acid}}{\overset{\displaystyle HO}{\underset{\displaystyle HO}{>}}C = O} \right] + 2\,HCl$$

$$\downarrow\; -H_2O$$

$$\underset{\text{Carbon dioxide}}{O = C = O}$$

Figure 1.3   The hydrolysis of phosgene.

Direct toxic agents, which do not undergo any preliminary enzymatic transformation, interact with more or less specific biological targets causing more or less serious structural modifications of these targets. On this basis, direct toxic agents may be grouped into two major categories:

Direct toxic agents that do not fundamentally modify their targets. As a result of its strong affinity for a specific target, referred to as a *receptor*, the xenobiotic will bind via reversible bonds of weak energy (hydrogen bonds, hydrophobic interactions, etc.). The reversibility of the xenobiotic's binding to its receptor is variable and may become practically irreversible if the affinity between the target and its ligand is very high. This is the case for certain toxins that bind very strongly to their membrane receptors and for natural products such as colchicine that bind almost irreversibly to tubulin (one of the proteins that constitutes the cellular cytoskeleton). These substances are notoriously toxic and are extremely difficult to displace from their respective targets.

The receptor capable of binding a direct xenobiotic may be of variable nature:

Membrane. Cellular biomembranes are constituted of phospholipids, most often unsaturated, and of proteins, many of which can serve as receptors (ionic channels, etc.).

Enzymatic. Many toxic xenobiotics act by binding to an enzyme and inhibiting it, sometimes irreversibly. Certain chemical substances, such as organophosphorus compounds or eserine (an alkaloid), specifically inhibit circulating enzymes (acetylcholinesterase, etc.). Other toxic agents selectively bind to essential intracellular enzymes. This is the case for very efficient acute toxic substances such as cyanides, azides, hydrogene sulfide, and so on.

Glycoproteins. Membrane or circulating glycoproteins (e.g., antibodies), involved in the mechanisms of immunity, may be the privileged targets of certain immunotoxic xenobiotics. Many strongly alkylating macromolecules such as simple aldehydes (formaldehyde, acrolein, glutaraldehyde, etc.), quinones, lactones (α-methylene-γ-lactones, etc.), interact with immunoproteins and may lead to delayed hypersensitivity reactions (contact allergies).

Water-soluble, intracellular proteins. Certain water-soluble cytosolic proteins implicated in the transport of endogenous substrates selectively bind xenobiotics, contributing to their cytotoxicity. This is the case for polyhalogenated aromatic or heterocyclic compounds such as polychlorinated biphenyls (PCBs), polychlorinated dibenzofurans (PCDFs), and polychlorinated dibenzodioxins (PCDDs), the tetrachloro derivative of which (TCDD) is more generally known under the name of dioxin.

Circulating carrier proteins. Circulating carrier proteins may selectively bind certain xenobiotics, thereby inhibiting their capacity for transport and sometimes leading to serious afflictions. Thus, the very strong binding of carbon

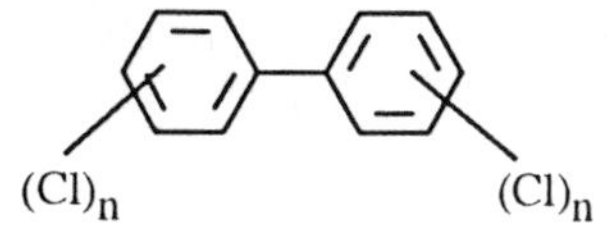

Polychlorinated biphenyls (PCBs)

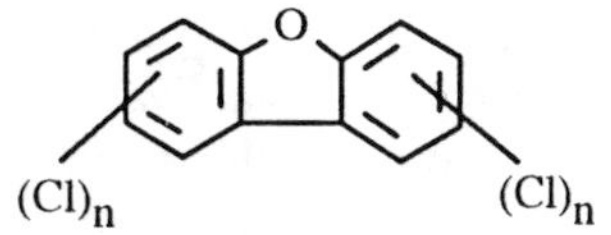

Polychlorinated dibenzofurans (PCDFs)

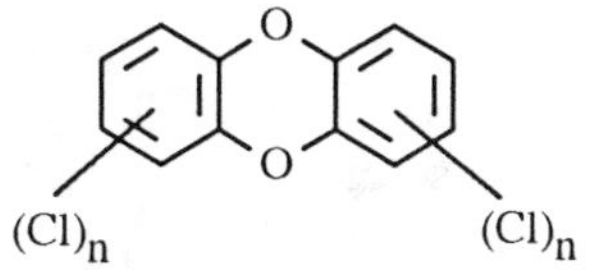

Polychlorinated dibenzodioxins (PCDDs)

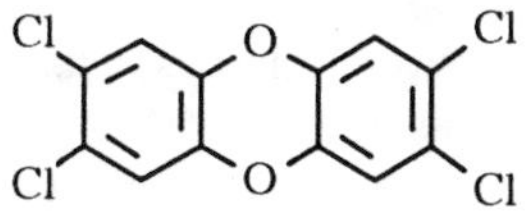

2, 3, 7, 8 - Tetrachlorodibenzo-p-dioxin
(TCDD) or Dioxin

monoxide to hemoglobin, the hemoprotein responsible for oxygen transport, quickly leads to symptoms of asphyxiation (cyanosis, anemia, etc.).

Intracellular cytoskeletal proteins. Tubulin, one of the constituent proteins of cytoskeletal microtubules, very strongly binds various toxic xenobiotics (heavy metals, colchicine, quinones, etc.). Similarly, certain neurofilament proteins selectively bind to 2,5-hexanedione, a peripheral neurotoxin that can cause debilitating polyneuritis.

Nucleic acids. Among the nucleic acids, DNA is a major target of many xenobiotics which modify it either by intercalation between bases (e.g., intercalating agents such as acridine dyes or ethidium bromide) or by direct alkylation, leading to mutations having more or less serious consequences (e.g., development of tumors, malformations, etc.).

Direct toxic agents that fundamentally modify their targets. These agents are generally highly reactive substances that either more or less completely destroy their biological targets (e.g., corrosive substances) or modify them by alkylation (direct alkylating agents) or oxidation (strong oxidizing agents).

Corrosive substances. Corrosive substances are highly reactive and provoke the brutal and irreversible destruction of the tissues with which they come into contact. Corrosive chemical compounds exhibit a wide variety of reactivity depending on their chemical structure, their molecular weight, their polarity, and their binding capacity. Their reactivity toward biological constituents may be associated with:

   –their acidity or basicity, that is, the degree of dissociation of their ionizable
     function
   –their oxidizing or reducing strength or potential
   –their dehydrating capacity, that is, the capacity to remove the elements of
     water from a given structure

     –their alkylating strength and the possibility of attaching themselves (as is or after metabolic activation) to a biomolecule
     –their affinity for cellular lipids
     –their physicochemical state

These substances primarily attack skin and mucosa (ocular, nasal, respiratory, digestive).

Liquids, gases, and solids are able to produce corrosive effects. Most liquids immediately attack tissues (e.g., strong mineral bases and oxidizing agents), but some, such as concentrated aqueous acids (HCl, HBr, HI, HF), act more slowly, leaving enough time to rinse them off in case of contact by splatter, thereby avoiding penetration. Note, however, that HF is a bone seeker and can cause serious injury if not properly treated. Gases, vapors, and microparticle aerosols ($<5\ \mu$m) mainly affect ocular and respiratory mucosa. Certain highly aggressive gases such as yperite (sulfur mustards) attack skin after variable latency periods. Solids react more slowly with tissues, but some are particularly dangerous to skin and mucosa — e.g., white phosphorus, dehydrating agents (phosphoric oxide), solid hydrides used as reducing agents in synthesis (lithium aluminum hydride).

Alkylating agents. Alkylating agents are organic compounds capable of introducing onto a given molecule a hydrocarbon group of the alkyl type (i.e., a group composed of carbon and hydrogen). They are considered to be electrophilic reagents, that is, electron-deficient entities. As a result, they will react with electron-rich substrates or nucleophiles. At the cellular level, alkylating agents react with proteins and nucleic acids, transforming these cellular constituents into substituted derivatives. These modifications prevent them from carrying out their normal functions.

These compounds often exhibit delayed toxicity, especially as regards the pulmonary system, which may appear several hours after contact. This is the justification for prolonged medical surveillance, often misunderstood because of its constraints.

Table 1.8 lists the principal types of direct toxic agents.

Indirect toxic agents (protoxic agents)

The majority of xenobiotic chemical products require prior enzymatic conversion by the organism before toxic effects appear. These are referred to as *indirect toxic agents* or *protoxic agents*. It is the reactive intermediates or the products of metabolism that are responsible for the appearance of the toxic process.

Most of the metabolic reactions of xenobiotics occur in the liver. However, metabolizing enzyme systems also exist in other organs (kidney, placenta, lungs, digestive tract, skin, nasal cavity). This can explain the selective toxicity of certain compounds. Figure 1.4 gives the principal types of toxic agents classed according to their reactivity.

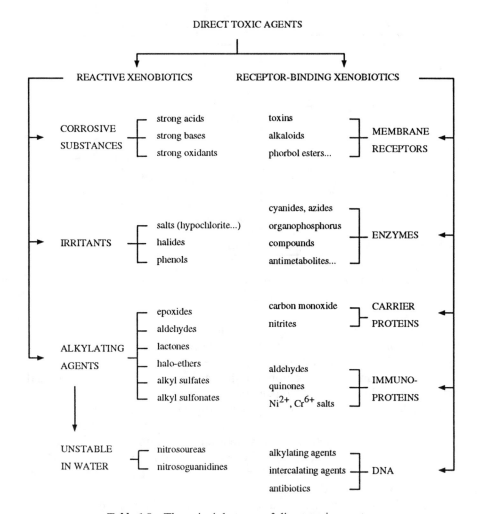

Table 1.8 The principle types of direct toxic agents.

Duality between detoxification and intoxication

Once inside an organism, an organic xenobiotic compound, if not eliminated as is, will be neutralized by metabolic enzymatic systems which will transform it into a more water-soluble compound. This hydrosolubilization generally requires two successive steps. The first step consists of the introduction of a polar functional group (alcohol, phenol, etc.) onto the molecule. The primary metabolite thus formed is generally not sufficiently water soluble, however. Thus, in a second step, referred to as *transfer* or *conjugation*, this primary metabolite adds onto the functional group introduced in the first step a small

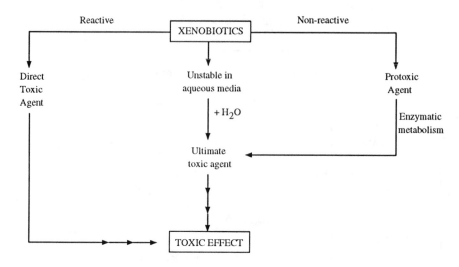

Figure 1.4   Principle types of reactivity-dependent toxic agents.

endogeneous molecule (sulfuric acid, glucuronic acid) able to form a water-soluble alkaline salt. It is this final metabolite that will be excreted, generally by way of the kidneys and urine.

The transformation of a xenobiotic into a more water-soluble compound occurs principally in the liver, our main "antipoison center."

In the hepatic (liver) cell, the major detoxifying enzymatic system is found in a complex membrane structure, the smooth endoplasmic reticulum.

The enzymes of the detoxifying system are mainly cytochrome P_{450} mono-oxygenases. Oxidations constitute the most common type of reaction for transforming lipophilic substances into water-soluble compounds (e.g., by hydroxylation of C–H, C–N or C–S bonds, oxidations, oxidations of double bonds, etc.). These enzymes are also found in lungs, kidneys, bone marrow, skin, and placenta.

Unfortunately, it is often during this beneficial metabolism that intermediates appear which can react either directly at the sites where they are formed or at sites to which they are transported. These electrophilic intermediates form irreversible covalent bonds with nucleophilic cellular macromolecules. Interactions with proteins lead to more or less serious necrosis, to immunological impairment, and the like, while interactions with nucleic acids (DNA) may provoke the appearance of mutations eventually followed by the formation of tumors (Figure 1.5).

As may be noted in Figure 1.5, it is the ratio between the efficiency of the detoxifying route (elimination of water-soluble metabolites) and the importance of the intoxication process (formation of nonspecific, irreversible bonds with various cellular targets) which is responsible for cellular degradation (necrosis, mutations, and, eventually, tumor formation).

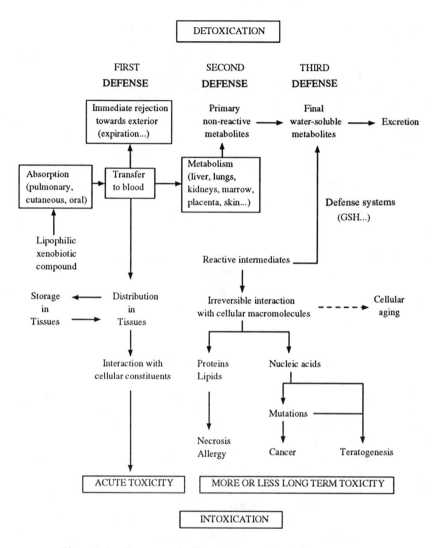

Figure 1.5   Outcome of a lipophilic xenobiotic in an organism.

1.3.2.2 Types of Toxicity According to Effects

Acute toxicity

This is the result of the administration of a single dose or of fractions of this dose over 24 hours.

In the work environment, there are numerous cases of acute toxicity whose symptoms do not appear before several hours, particularly if the toxic agent is likely to cause acute pulmonary edema (e.g., gases and vapors easily hydrolyzed in the lungs: Section 1.3.4.2). This fact amply justifies prolonged observations, generally for 48 hours, in the emergency or recovery wards of a hospital.

Estimation of acute toxicity is based on the determination of the dose of a substance causing death (lethal dose or LD) or a particular anomaly such as a nervous disorder, or altered blood compositions (effective dose or ED). For operational reasons, the LD_{50} and ED_{50}, representing mortality or the appearance of an expected symptom in half of the tested population, have been defined. Since doses are difficult to evaluate in the case of intoxication by inhalation, the concentrations are measured instead (LC_{50} and EC_{50}).

The estimation of the acute toxicity of a given substance gives indications concerning the probable effects of a massive exposure to humans in the case of exceptional situations (accidents, suicides, etc.).

Subacute toxicity

Subacute toxicity is the result of the repeated administration of a substance over a period of 14 days to 3 months. Information concerning the potential toxic effects of exposure over a limited time is thus obtained, as well as information concerning the affected target organ.

Long-term or chronic toxicity

Long-term toxicity is the result of repeated and frequent exposures to small or very small quantities of toxic substance over periods ranging from a few months to several years. Intoxication is the result of either the accumulation of the toxic agent in an organ, as with lead, or of the additive effects of several agents, as with carcinogens.

For a particular toxic agent, the symptoms of intoxication can vary depending on the physical state of the agent and on whether the intoxication is acute or long term. Thus, cadmium oxide (CdO), inhaled in the form of roughly ground powder, causes yellowing of teeth, while as a fine powder causes various physiological problems (digestive, renal, pulmonary). Inhaled in the form of smoke, however, it provokes acute pulmonary edema. Acute intoxications with benzene lead to loss of consciousness, coma, and death due to depression of cerebral activity; in long-term intoxication, alteration of blood composition can lead to death due to aplastic anemia or leukemia. These two examples demonstrate the arbitrary character of these classifications and the complexity of these modes of intoxication.

Remark: Although the predictive value of toxicity studies in animals is limited, particularly regarding extrapolations that can be made to humans, it must be recognized that, in view of present knowledge, animal experimentation constitutes an indispensable scientific prerequisite in defining the toxic risks of chemical substances. Nevertheless, animal experimentation cannot be dissociated from the quality of the conditions under which it is conducted. For this reason, the European Economic Community (EEC) and many other countries, in particular the United States, have adopted regulations such that laboratory animals may benefit from the best possible protection and that all unnecessary suffering or unjustified sacrifice may be avoided. Research is presently being conducted in order to develop alternative or complementary in vitro techniques, in particular using cell cultures, which, with time, should allow a decrease in the number of animals used in experimentation, especially in screening. This in vitro testing may also provide faster results from shorter-term testing.

1.3.2.3 The Modes of Expression of Toxicity

The majority of toxic effects correspond to more or less specific interactions with vital cellular constituents and, in particular, with macromolecules such as proteins, unsaturated lipids, and nucleic acids. Figure 1.6 summarizes these interactions.

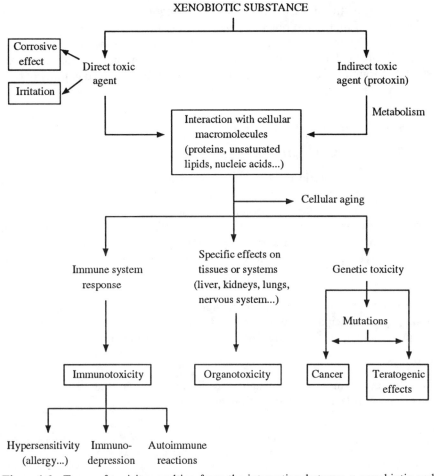

Figure 1.6 Types of toxicity resulting from the interaction between a xenobiotic and the cellular constituents of the human organism.

The vital macromolecules so modified can no longer assure their functions, resulting in cell degradation. Depending on the type of macromolecule attacked by these reactive intermediates, toxic effects will be observed over variable long-term periods. The degradation of biomembrane lipoproteins will lead to cell necrosis.

Similarly, xenobiotic substances, by attaching themselves to membrane-bound or circulating proteins, stimulate the immune system to produce antibodies. These, by binding to sensitized cells, liberate various mediators (histamine, prostaglandins, leukotrienes, platelet activating factor, etc.) which are responsible for inflammatory and allergic phenomena. Inflammation always begins with vascular symptoms—for example, redness, swelling, impression of heat on the skin—before the advent of pain.

The modification of DNA may result in a mutation (among other things) which, if it is not eliminated or is improperly repaired, will lead to the initiation of tumor growth. If DNA is attacked during the embryonic period, the appearance of malformations in progeny is possible.

In summary, the manifestation of a long-term toxic effect varies as a function of the biological target attained (cell necrosis, immunotoxic or genotoxic effects). Though difficult to estimate numerically, the prolonged contact of cells with toxic chemical substances undeniably accelerates their aging.

1.3.3 Modes of Intoxication

1.3.3.1 Target Organs

While many xenobiotic chemical compounds, such as corrosive agents, act nonspecifically, others interact specifically with precise cellular targets. Specific lesions of various tissues and organs may then be observed (organotoxicity). Tissues and organs by which chemical agents enter (skin, lung, digestive tract, etc.) or leave (liver, kidney, etc.) are particularly exposed to the noxious effects of xenobiotics (Figure 1.5). Only particularly exposed organs will be discussed here.

Liver

Xenobiotic substances are mainly metabolized in the liver and, as such, this organ is the principal target for many of these agents.* Table 1.9 summarizes the principal hepatotoxic substances encountered in the laboratory.

Kidneys

The kidneys, which regulate salt and water levels in the organism, play an essential role in the elimination of numerous substances. In the course

---

*Hepatitis may be classified as toxic or immunoallergic. Toxic hepatitis, the most frequent, is due to relatively unstable toxins which can react only at the site where they are formed. The mechanisms vary considerably (necrosis, cirrhosis, immune response) and are still incompletely understood (covalent bonds to proteins and/or nucleic acids; peroxidation of membrane lipids; depletion of hepatic glutathion; imbalance of certain ions such as calcium; etc.).

CHEMICAL FAMILY	COMPOUND
Halogenated organic derivatives	chloroform, iodoform, carbon tetrachloride, tetrachloroethane, vinyl chloride, trichloroethylene, perchloroethylene, chlorobenzene, bromobenzene, 1,2-dibromoethane, 1,2-dichloroethane, halothane
Nitrogenous organic derivatives	2-nitropropane, dimethylnitrosamine, hydrazines, urethane, galactosamine
Inorganic substances	white phosphorus, beryllium, selenium

Table 1.9   Principal hepatotoxic substances.

of acute intoxications with, for example, lead, cadmium, chromium or uranium, the renal tubules are most often affected, leading to nephritis, a serious renal deficiency. Over a long period, however, mercury and halogenated solvents such as chloroform, carbon tetrachloride and tri-chloroethylene, provoke lesions of the glomerulus in particular, apparently by immunological mechanisms.

In the laboratory, long-term renal intoxications provoked by mercury, and especially its vapors, as well as by the halogenated solvents cited earlier, should be particularly guarded against.

Ethylene glycol is especially toxic by ingestion (lethal dose in humans: 1.4 ml/kg).

Lungs

The respiratory system, the principal route of penetration by toxic agents in the laboratory, is particularly exposed to intoxication. The inhalation of corrosive substances (see Section 1.3.4.3) or of irritants (see Section 1.3.4.4) can lead to acute pulmonary lesions, most often in the form of edemas and sometimes with delay.

Many irritating gases and vapors such as ozone, nitrogen oxides, formaldehyde or pulverized substances such as asbestos, can lead over long periods to serious pulmonary diseases such as asthma or bronchial cancers. Cigarette smoke strongly potentiates these effects.

The nervous system

In the nervous system (rich in lipids and, especially, in phospholipids), lipophilic (or fat-soluble) xenobiotic molecules may perturb the transmission of neuronal signals, provoking excitation ("drunkenness"), depression (anesthesia, narcosis), and sometimes death. Certain toxins that can accumulate in the nervous system produce long-term diseases such as encephalitis (mercury, bismuth, toluene, etc.) or polyneuritis (hexane, acrylamide, carbon disulfide, etc.). A few neurotoxins exert their destructive effects at specific sites. Thus, methanol attacks the optic nerve, leading to blindness, while the action of manganese on the *corpus striatum* in the brain provokes neuronal degeneration of the Parkinsonian type. The accumulation of even small doses of certain toxins (e.g., heavy metals, halogenated solvents) in the nervous system may contribute to various psychosomatic problems wrongly attributed to constraints of the laboratory working environment.

Table 1.10 summarizes the principal neurotoxic chemical substances.

Genotoxicity

The interaction of chemical substances with nucleic acids, and particularly with deoxyribonucleic acid (DNA) contained in the chromosomes, is particularly serious. This genetic toxicity is involved in the development of tumors and in the appearance of teratogenic effects in progeny (malformations).

Among xenobiotic chemical substances likely to interact with cellular macromolecules, only three classes of compounds, whose insidious effects are to be guarded against during the course of laboratory work, will be considered here:

−allergenic substances
−carcinogenic substances
−teratogenic substances

1.3.3.2 Allergenic Substances

The contact of certain chemical substances with skin or mucosa or their introduction into the organism may set off an abnormal reaction of the immune system. Such substances are referred to as *allergenic* while the problems that arise as a result of this response of the immune system are generally referred to as *allergies*. Allergies may be considered as both direct and indirect toxicity.

Substances may penetrate into the organism by percutaneous (skin, mucosa), respiratory, or oral routes and can accordingly act on different tissues or organs (skin, eye or nasal tissues, lungs, liver, kidneys, etc.). A variety of problems may arise, depending on the target (contact dermatitis [eczema], conjunctivitis, rhinitis, asthma, bronchitis, etc.).

A first contact with an allergenic substance does not generally lead to observable effects due to the necessary incubation period, except if the substance is also a primary irritant (see Section 1.3.4.4).

CHEMICAL SUBSTANCE	NERVOUS SYSTEM TARGET
Mercury Aluminum Glutamate	Neuronal cell body
Acrylamide Hexane 2-Hexanol 2,5-Hexanedione Methyl butyl ketone Carbon disulfide	Axon
Hexachlorophene Isoniazid Ethidium bromide Lead salts Alkaline dichloroacetates	Myelin
Lead salts Cadmium salts	Blood vessels irrigating the nervous system
Carbon monoxide Alkaline cyanides Alkaline azides	Anoxia (decrease in the quantity of oxygen which blood can distribute to tissues)
Organophosphorus compounds - diisopropylfluorophosphate (DFP) - malathion - parathion - fenthion - diazinon Carbamates - carbaryl - aldicarb	Acetylcholinesterase

Table 1.10   The principal neurotoxic chemical substances.

Allergic reactions, also referred to as hypersensitivity, call into play various mechanisms of the immune response. These include humoral immunity as shown by the appearance of circulating antibodies (immunoglobulins) and cellular immunity indicated by formation of membrane-bound antibodies.

Most of the allergenic substances encountered in the laboratory give rise to contact allergies corresponding to type IV allergies or delayed hypersensitivity (based on the Gell and Combs classification). The chemical substance responsible for contact allergy, generally of low molecular weight (between 500 and 1,000 g/mole) is called a *hapten* or *incomplete antigen*. In most cases, this agent is not by itself sufficient to produce an allergic reaction. In order for it to be recognized by the organism, the agent must be coupled to a "carrier" molecule which is generally a protein. This process is referred to as *haptenization*. This explains why in contact allergies, symptoms do not appear on the first exposure to the chemical agent. Once sensitization is achieved, however, succeeding contact with the same agent may produce symptoms within 24–48 hours. Symptoms may also appear 7–10 days after exposure but, in many cases, an allergic reaction develops after several years of continuous exposure to low concentrations of allergen.

When skin comes into contact with an allergenic substance, eczema is produced which may spread over a variable area. It is generally characterized by a reddening of the skin (erythema), a swelling of cutaneous tissues (edema), and the appearance of itching and blisters. When mucosa of the respiratory tract come into contact with an allergenic substance, bronchoconstriction develops, as manifested by asthmatic attacks, rhinitis, and so on.

Three phases in the cellular mechanism of contact allergy may be distinguished:

–Induction, in which the organism comes into contact with the hapten (i.e. the aggressing chemical substance) for the first time. The hapten is coupled to a protein and the complex so formed (the allergen or complete antigen) is digested by a Langerhans cell. The latter then presents the allergen to T lymphocytes (white blood cells specialized in defense against foreign intrusions). During this step, "information" is transmitted from the Langerhans cell to the T lymphocytes.

–Proliferation, which also occurs on the organism's first exposure to an allergen. The "informed" T lymphocyte migrates toward the closest ganglion, multiplies, and produces subpopulations of "auxiliary" T lymphocytes and "memory" T lymphocytes.

–Reaction which occurs only on renewed contact with an allergen. Clinical symptoms appear, marking the onset of the contact allergy. The hapten, having penetrated into the skin or the mucosa, is transformed into an antigen and then bound to a Langerhans cell which, in turn, presents it to an "informed" or immunocompetent T lymphocyte. The latter proliferates in the nearest ganglion and the effector cells which result are responsible for the inflammatory reaction that occurs.

It has been demonstated that most cutaneous or respiratory allergenic chemical substances have electrophilic properties (i.e., they are deficient in electrons). They thus react directly with amines (NH_2), thiols (SH), alcohols, and phenols (OH). Some long-chained hydrocarbons may act by a lipophilic type of mechanism. These haptens, having both a hydrophilic moiety (soluble in water) and a hydrophobic (or lipophilic) moiety (the hydrocarbon chain, soluble in organic solvents), can insert themselves into cell membranes.

Certain compounds, such as formaldehyde or benzoquinone, are direct electrophiles and are thus direct toxic agents. Others, such as *p*-phenyl enediamine, must first be metabolized in vivo (i.e., oxidation to *p*-benzo-quinone) in order to become veritable electrophiles. These are referred to as *prohaptens* or *protoxins*. Because of this, different molecules which can be metabolized into the same hapten may give crossed allergies. This is the case for hydroquinone and *p*-phenylenediamine which are both metab-olized into *p*-benzoquinone. Hydrolysis and esterification are other in vivo metabolic transformations which can cause innocuous substances to become electrophilic haptens.

Certain chemical substances are only active in the presence of light. Several aromatic examples of these photoallergens are given in Table 1.11.

Tetrachlorosalicylanilide	Phenothiazines
Trichlorosalicylanilide	Sulfamides
Tribromosalicylanilide	Quinine
Trichlorocarbonilide	p-Aminobenzoic acid
Fentrichlor	

Table 1.11   Some aromatic contact photoallergens.

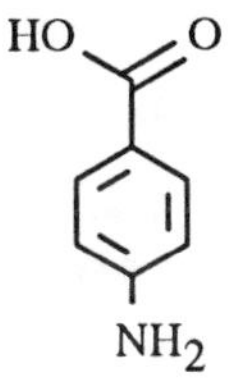

p-Aminobenzoic acid 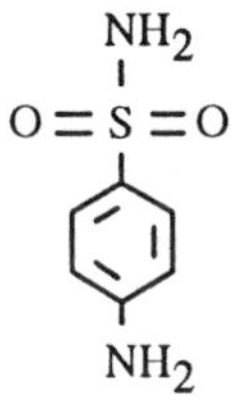   Sulfanilamide 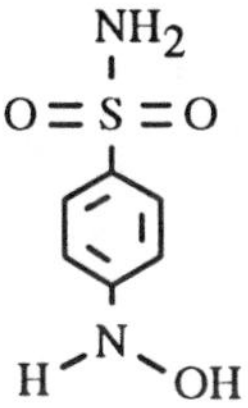   p-Hydroxyaminobenzene
sulfonamide

Numerous allergenic substances are encountered in the chemical laboratory, among which the following may be cited:

–Aliphatic amines such as ethylenediamine and alicyclic amines such as piperazine;

$H_2N\text{-}CH_2\text{-}CH_2\text{-}NH_2$

Ethylene diamine

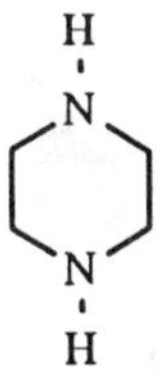

Piperazine

–Aromatic amines, particularly those in which the primary amine is situated in a *para* position as in *p*-phenylenediamine, *p*-aminophenol, and *p*-aminobenzoic acid. The substitution may also sometimes be in the *ortho* position as in *o*-phenylenediamine;

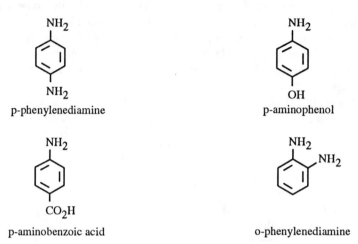

p-phenylenediamine

p-aminophenol

p-aminobenzoic acid

o-phenylenediamine

–Hydrazines such as hydrazine hydrate or methylhydrazine;

$H_2N\text{-}NH_2 \cdot H_2O$

CH$_3$-NH-NH$_2$

hydrazine hydrate

methylhydrazine

–Quinones such as *o*- and *p*-benzoquinone and their corresponding diphenol precursors;

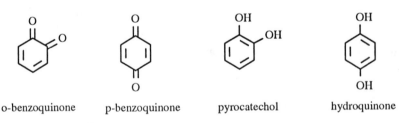

o-benzoquinone   p-benzoquinone   pyrocatechol   hydroquinone

–Aldehydes such as formaldehyde (formol) or acrolein;

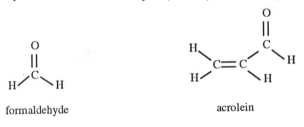

formaldehyde

acrolein

−Acrylic derivatives such as acrylic acid, acrylates, methacrylates;

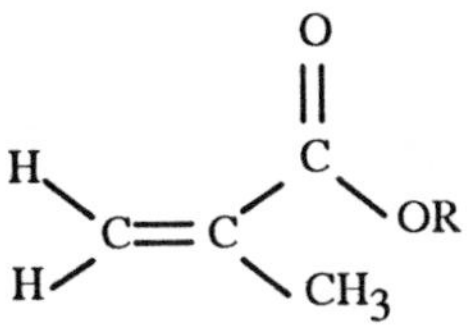

methlacrylate 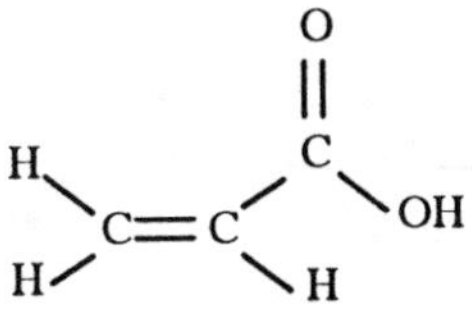   acrylic acid 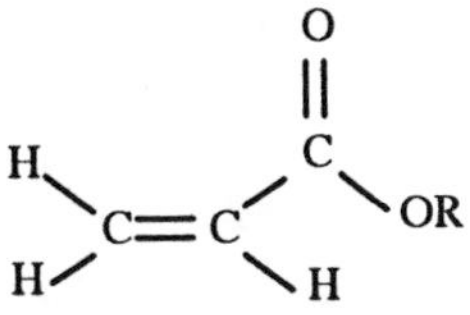   acrylate

Hexavalent chromium salts such as the chromates and dichromates found in sulfochromic mixtures: $Na_2Cr_2O_4$ sodium chromate, $K_2Cr_2O_7$ potassium dichromate;
−Nickel and its salts: Ni^0 nickel, Ni^{2+} divalent nickel salts.

In biological and biochemical laboratories, a variety of other allergenic substances may be handled:

−2,4-Dinitrofluorobenzene (DNFB) and sometimes 2,4-dinitrochlorobenzene (DNCB), compounds for which contact with less than 5 mg provokes strong allergic reactions;

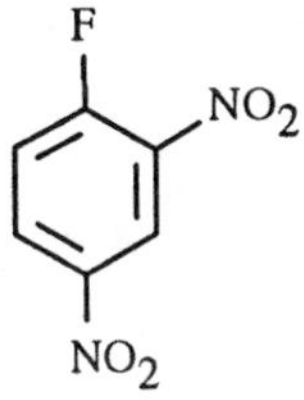

DNFB 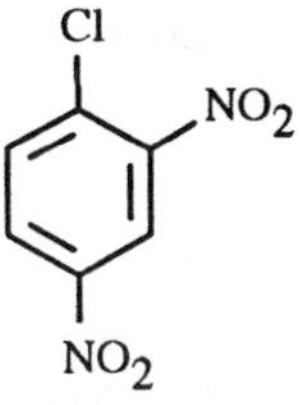   DNCB

−Glutaraldehyde;

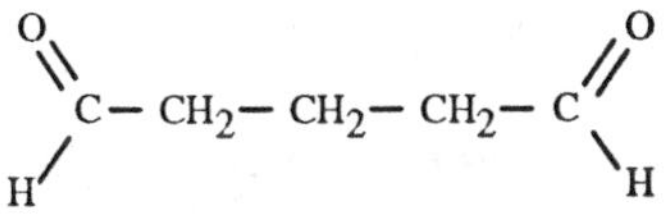

Glutaraldehyde

−Picric acid (a constituent of Bouin's liquid, a fixing agent used in histology);

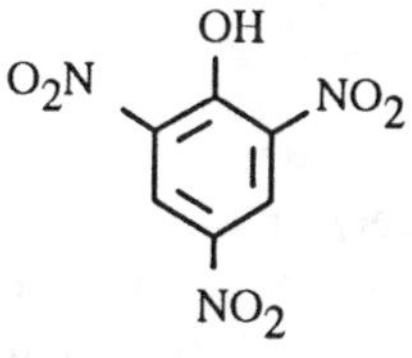

Picric acid

–Epoxy resins used for the preparation of histological samples.

General pattern of an epoxy resin

Finally, certain plants and various biologically active substances (penicillins neomycins, etc.) can present allergenic properties. Exposure to sunlight may also initiate photoallergic or phototoxic reactions.

Protection against allergenic substances calls for individual preventive measures such as the wearing of goggles, gloves, masks, and protective clothing. In addition to rigorous personal hygiene, collective preventive measures, such as awareness of risks, proper ventilation, and cleanliness of work areas, are also necessary.

It should be noted that certain protective gloves may contain allergenic substances such as 2-mercaptobenzothiazole (MBT), tetramethylthiuram disulfide (TMTD), and so on.

2-Mercaptobenzothiazole (MBT)   Tetramethylthiuram disulfide (TMTD)

The principal allergenic substances encountered in the laboratory are listed in Table 1.12.

1.3.3.3 Carcinogenic Substances

Mechanism of carcinogenesis

At present, it is generally admitted that cancers are the result of a series of events, many of which are still largely unknown. DNA (the source of genetic information), contained in the chromosomes of the cell nucleus, is apparently the privileged target of carcinogenic agents (chemical substances, ionizing radiation, viruses). According to this "genetic" hypothesis, the transformation

CHEMICAL FAMILY	EXAMPLES OF COMPOUNDS	ALLERGIC REACTION
Beryllium	Be, BeO, BeF_2	Dermatitis, pulmonary berylliosis
Chromium	Cr^{3+}, Cr^{6+} *	Eczematic dermatitis, asthma
Cobalt	Co, Co^{2+}, Co^{3+}	Dermatitis
Mercury	Hg, $HgCl_2$	Glomerulo-nephritis
Nickel	Ni, Ni^{2+}	Dermatitis
Osmium	Osmium tetraoxide (OsO_4)	Eczema, conjunctivitis, rhinitis
Platinum	Hexachloroplatinic acid ($PtCl_6H_2$), chloroplatinates	Dermatitis, asthma
Aldehydes	Formaldehyde, furfural, acrolein, glutaraldehyde	Dermatitis, asthma
Acid anhydrides	Maleic, phthalic and trimellitic anhydrides	Asthma, rhinitis
Unsaturated esters	Acrylates, methacrylates	Dermatitis
Epoxides	Ethylene oxide, epichlorohydrin, glucidic ethers, epoxy resins	Asthma, rhinitis, dermatitis
Phenols	p-t-Butylphenol, p-t-butylcatechol, hydroquinone	Dermatitis, skin depigmentation
Quinones	p-Benzoquinone	Dermatitis
Polynitrated haloarenes	2,4-Dinitrofluorobenzene, 2,4-Dinitrochlorobenzene picryl chloride	Dermatitis
Aliphatic amines	Methylamines, ethylamines, butylamines, cyclohexyl-amine, ethylenediamine	Dermatitis
Cyclic amines	Piperazine	Dermatitis
Aromatic amines	p-Phenylenediamine, p-diaminotoluene, p-aminophenol	Dermatitis
Hydrazines	Hydrazine, phenylhydrazine	Dermatitis
Isocyanates	Toluene diisocyanates	Dermatitis, conjunctivitis, asthma

Table 1.12   Principal Allergenic Substances.

*Chromium (III) salts do not penetrate skin and thus cannot express their allergenic (hapten) potential. On the other hand, chromium (VI) penetrates cutaneous barriers, after which it is reduced to chromium (III) enabling formation of antigenic complexes.

of a normal cell into a cancerous cell occurs via two essential steps: initiation followed by promotion. Initiation occurs at the DNA level and can result in one or more irreversible mutations. After a more or less extended latency period, promotion leads to a disorganization of cellular growth and differentiation with the appearance of cancerous (transformed) cells. According to this unitary theory, the carcinogenic process begins with structural alterations (mutations or transpositions) of a few genes (a unit of genetic information composed of a more or less long sequence of nucleotides which compose the DNA double helix) occurring randomly. These DNA modifications lead, in turn, to structural modifications of different portions of the genome, provoking activation of the expression of a previously latent "cancer gene." Activation of this "cancer gene" leads to the synthesis of a specific protein, thereby modifying the normal activity of the cell with respect to its environment.

In general, the same substance could have both initiation and promotion functions (a complete carcinogen). This merely requires that the substance act for a sufficiently long period of time. This is the case for the majority of the carcinogens listed in the following figures.

As a consequence, a chemical substance should not only be mutagenic, but should also have a promoting action in order to be considered carcinogenic. This explains why powerful mutagens, such as sodium azide (NaN_3) or hydroxylamine (NH_2OH), are not carcinogenic. Inversely, certain promoter substances may, by themselves, possess carcinogenic activity without being mutagenic. This is the case, for example, of inorganic substances such as asbestos, arsenic, and the like. It appears quite clearly that promoter activity must have a very great importance in the genesis of cancers; unfortunately, this type of activity cannot as yet be easily determined for chemical substances.

Mechanism of action of a particular type of carcinogenic substance, the alkylating agents.   Alkylating agents are organic compounds capable of introducing an alkyl group (a simple hydrocarbon chain) onto a given molecule. By extension, the term *alkylation* is also used for reactions in which aromatic groups (arylation) or carboxylic acid acyl groups (acylation) are introduced.

Thus, alkylation consists of introducing an alkyl chain R, that is, a moiety composed of carbon and hydrogen, onto a molecule M. Depending on the number of carbon atoms introduced, the terms *methylation* (addition of a methyl group, $-CH_3$), *ethylation* (addition of an ethyl group, $-CH_2CH_3$), and so on, may be employed.

$$M + R \rightarrow M - R \text{ (alkylated molecule)}$$

Numerous reactions allow creation of a new covalent bond between an alkyl group R and a carbon atom or a heteroelement (nitrogen, sulfur, oxygen, phosphorus, metal) of a substrate. For example, certain metals (mercury, lead, tin, bismuth, etc.) and metalloids (selenium, etc.) can undergo biomethylations in some environments (e.g., by microorganisms, such as those found in the intestine), forming methylated derivatives that are more toxic than the starting metallic compounds.

Among the many alkylation reactions, only those which involve nucleophilic substitution (S_N) will be considered. These proceed via displacement of a functional group X bound to an alkyl group R by a nucleophilic substrate M:⁻ (or M:), that is, a substrate rich in electrons (Figure 1.7).

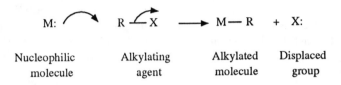

| Nucleophilic | Alkylating | Alkylated | Displaced |
| molecule | agent | molecule | group |

Figure 1.7   A nucleophilic substitution reaction.

In this type of reaction, the leaving group X may be a halide (chloride, bromide, iodide, etc.), an ester (sulfate, sulfonate, phosphate, etc.), a diazonium salt (which may form in vivo by activation of nitrosoamines, nitrosoamides, nitrosoguanidines, etc.) and so on.

As a general rule, alkylating agents are considered to be electrophilic reagents, that is, electron-deficient entities which, as a result, react with electron-rich substrates or nucleophiles (Figure 1.8).

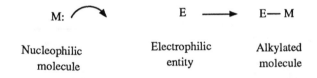

| Nucleophilic | Electrophilic | Alkylated |
| molecule | entity | molecule |

Figure 1.8   Reaction between a nucleophilic molecule and an electrophilic entity.

At the cellular level, proteins and nucleic acids possess many more nucleophilic sites than electrophilic sites. It is thus with these cell constituents that alkylating agents react, transforming them into substituted derivatives such that they can no longer carry out their normal functions.

The irreversible degradation of proteins may lead to more or less serious tissue necrosis. Thus, dialkyl sulfates and alkyl fluorosulfonates are highly corrosive to skin, eyes, and the respiratory tract (acute pulmonary edema). Other alkylating agents are eye (lachrimators) and skin irritants.

It appears that the privileged cellular targets of alkylating agents are the nucleic acids and, in particular, the deoxyribonucleic acids (DNA) which make up the genes, that is, the genetic program.

The nucleic acids are giant molecules present in all living cells. They are formed by the linkage of simpler units, the nucleotides, each of which is composed of a nitrogenated heterocyclic base derived from a pyrimidine (cytosine, uracil, and thymine) or a purine (adenine and guanine), a five-carbon sugar (pentose), and a phosphate group. In ribonucleic acids (RNA), the pentose moiety is ribose while in deoxyribonucleic acids (DNA), it is deoxyribose. The sequence formed by the different DNA bases allows transmission of genetic information.

Human DNA contains approximately 12 billion base-pairs distributed between two identical groups of 23 chromosomes. Most of the DNA present in the cell nucleus is intimately associated with proteins called *histones*, forming a compact structure, the *chromatin*. Uncoiled, DNA measures more than 1 m in length. In mammals, chromosomal DNA forms between 50,000 and 100,000 genes, the latter being the units of genetic information.

In living systems, it is important to realize that DNA is emetabolically stable and it is not regenerated during the course of a cell's life span. RNA, however, has a variable life expectancy (a half-life of 3 minutes for messenger RNA).

According to the Watson and Crick model (1953), DNA is constituted of two complementary polynucleotide chains which wind along a common axis forming a double helix. This double helical, three-dimensional structure is not the only possible one for DNA and other forms are known to exist. The essential role played by DNA in the transmission of genetic information may be schematically summarized as a series of three reactions (Figure 1.9).

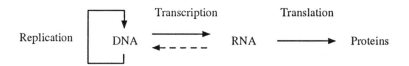

Figure 1.9   The transcription of genetic information.

In the initial duplication (or replication) step, one DNA molecule gives rise to two daughter DNA molecules, identical to the starting molecule, thereby ensuring the faithful reproduction of information. In the second transcription step, DNA transfers its information to a particular type of RNA molecule, messenger RNA (mRNA), whose nucleotide sequence constitutes the genetic code. In the final translation step, the mRNA, via organelles in the cytoplasm, translates its message into specific proteins. This synthesis is accomplished with the help of transfer RNA (tRNA) which adds the amino acids which it carries to the growing protein chain. The mRNA thus allows translation of a coded nucleotide message (the codons) into an amino acid message, thereby assuring the biosynthesis of the proteins which are the basic cell constituents (cell wall, carrier, enzymatic proteins, etc.).

In reality, this sequence of genetic information transfer is much more complex than Figure 1.9 indicates.

Alterations in DNA structure due to alkylation of purine (guanine, adenine) or pyrimidine (cytosine, etc.) bases may sometimes produce irreversible transformations (Figure 1.10).

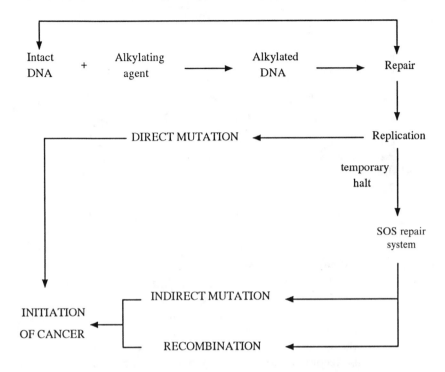

Figure 1.10 Development of tumors starting from DNA alkylation.

Though DNA alkylation can sometimes cause irreversible lesions which ultimately lead to cell death, the alkylation of bases at certain sites more often produces only slight modifications of the DNA which, after repairs, continues to replicate. The coded message of the newly formed DNA molecule will be erroneous. This is referred to as a *direct mutation.*

All living cells possess enzymatic systems capable of repairing lesions which appear in DNA. Some of these enzymatic systems completely restore the genetic information by cutting out the lesioned fragment of DNA (excision) and replacing it by an intact DNA sequence (repair), as symbolized in Figure 1.11.

Simple alkylating agents such as those producing methylations and ethylations have their own repair system. This consists of either direct excision of the alkylated bases by an N-glycosylase or the intervention of an alkyl transferase, the ADA protein.

The "cut and patch" system of repair, ineffective in repairing methylated or ethylated bases, intervenes in the case of lesions which cause major deformations of DNA (e.g., those caused by UV irradiation, aflatoxins, benzo(a)pyrene, psoralens, etc.).

Certain alkylated bases constitute a signal which sets off a general repair system familiarly known as "SOS repairs." This repair system may either serve to reconstitute the DNA's double helix by a recombination process or it may

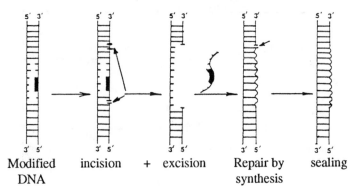

Modified   incision   +   excision   Repair by   sealing
DNA   synthesis

Figure 1.11 Repair of modified DNA by excision and resynthesis.

allow DNA replication while ignoring the errors introduced in the coded message. As a result, the DNA genetic code is modified and mutations appear. This case is referred to as *indirect mutation* (Figure 1.10).

According to the so-called genetic theory, a very close relationship exists between *mutagenesis* and *carcinogenesis*. A mutation constitutes an essential initiation step in the complex process whereby a normal cell is transformed into a malignant cell. Recombination can also play an important role in the initiation of carcinogenesis (Figure 1.10).

While this initial alteration of DNA may be considered a major event in the development of the tumoral process, even less well-understood factors must intervene afterwards to transform a normal cell into a malignant one.

It is recognized that cancer is generally due to a single cell line which, upon modification, eludes the normal controls regulating growth in the organism.

Based on an experimental model of skin cancer initiated by chemical substances in mice, three steps in the development of a tumoral process may be distinguished:

–an *initiation step* in which, under the influence of a carcinogenic agent (either directly or after its enzymatic activation), a normal cell is irreversibly transformed into a potentially malignant cell
–a *promotion step* in which a transformed cell proliferates under the influence of various endogenous factors (genetic, hormonal, or immunological), leading to a benign tumor (papilloma)
–a *progression step* in which cell proliferation results in formation of a malignant tumor (carcinoma).

In contrast to initiation, the ulterior steps are considered to be reversible. This three-step model, described by E. Hecker,* may be applied to other types

---

*In "Models, Mechanisms and Etiology of Tumour Promotion." IARC Scientific Publications, no. 56, IARC, Lyon (1984), pp. 441-63.

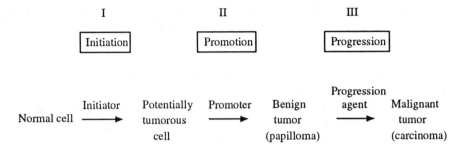

Figure 1.12   The different steps of experimental carcinogenesis in the mouse skin cancer model.

of cancer (Figure 1.12). According to some epidemiological studies, this model is also applicable to various human cancers.

The unitary theory of cancer.   Many agents are known to be capable of triggering cancers: viruses, ionizing (X-rays) and nonionizing (ultraviolet) radiation, hormones (sex hormones), food, xenobiotic chemical substances (foreign matter), and, of course, the association of two or more of these diverse causes.

The demonstration that viruses play a role in the development of cancers dates from 1911 when the Englishman P. Rous isolated the virus responsible for conjunctive tissue cancer in chickens (*Rous sarcoma*). This is in fact a *retrovirus* whose genome consists of RNA and which requires that an enzyme, reverse transcriptase (or retropolymerase), transform it into DNA before it can become integrated in the infected cell's chromosomes.

In 1976, in California, the Frenchman Dominique Stehelin identified the gene responsible for this cancerous transformation, the Sarc gene (named after the sarcoma which it produces). This was the first demonstration of a *cancer* gene or oncogene (V-onc). D. Stehelin also showed that normal animal cells contain a similar gene called a *proto-oncogene* (C-onc). Cellular proto-oncogenes are the source of viral oncogenes responsible for cancers. Apparently, a viral oncogene acquires a cancer gene by error as a result of random genetic recombination between its genome and that of the cell which it infects. Up to 40 oncogenes and their corresponding proto-oncogenes have presently been detected in animals.

In 1982, two American groups, that of R. Weinberg (MIT) and that of M. Barbacid (NCI) showed that a chemical substance (N-methylnitrosourea, an alkylating agent) can transform a dormant proto-oncogene into an active oncogene as a result of a simple modification of a single base among the 6,000 pairs of consecutive nucleic bases. These scientists demonstrated that, in human bladder cancer cells, the oncogene responsible for this cancer is in every way identical to the proto-oncogene except that a guanine (GGC codon) of the proto-oncogene is replaced by a thymine (GTC codon) in the oncogene. This unique modification provokes a change in the conformation of the synthesized protein (the SRC P21 protein) which now contains a valine residue in place of the glycine residue of the normal protein. This oncogenic protein acquires new properties which considerably modifies cell metabolism and leads to cancerous growth.

In summary, it is now generally acknowledged that the induction of cancer in a cell may be programmed by its own genes, the proto-oncogenes. These genes, under normal conditions, are repressed and dormant, but they may be activated to oncogenes by a variety of factors (viral, physical, or chemical). A single point mutation of a nucleic base may lead to the transformation of an inactive proto-oncogene into an oncogene capable of initiating cancer.

The activation of proto-oncogenes seems to be the general mechanism involved in the initiation step of carcinogenesis. A unitary theory of carcinogenesis by viral, physical, or chemical agents can thus be formulated. Nevertheless, the mechanisms of the steps succeeding initiation, that is, promotion followed by progression, are at the moment very poorly understood.

As a result of animal experiments (rats, mice), many alkylating agents have been shown to be carcinogenic (Table 1.13).

Depending on the route of administration, the localization of tumors may vary (fibrous tissue sarcoma after subcutaneous injection; lung cancer after repeated inhalation, etc.). Epidemiological studies concerning the carcinogenic potency of alkylating agents in humans are scarce. Only bis(chloromethyl) ether and chloromethyl methyl ether are known to be responsible for lung cancers. At least a dozen cases of bronchial cancers have been recently described in workers exposed to dimethyl sulfate. By simple analogy, it is possible that other alkylating agents known to be mutagenic and carcinogenic

Dimethyl sulfate	Diethyl sulfate
Methyl methanesulfonate	Ethyl methanesulfonate
Methyl iodide*	
n-Propyl iodide*	Isopropyl iodide*
n-Butyl iodide*	s-Butyl iodide*
Isobutyl bromide*	s-Butyl bromide
t-Butyl bromide*	
s-Butyl chloride	t-Butyl chloride*
Benzyl chloride	
Chloromethyl methyl ether	
Bis(chloromethyl) ether	Bis(chloroethyl) ether
Diazomethane	
1,2-Dibromoethane	
1,2-Dibromo-3-chloropropane	
1,2-Dichloroethane	
1,2-Dichloropropane	

*Alkylating agents whose intraperitoneal injection in mice causes a significant increase in lung cancer.

Table 1.13 Principal alkylating agents used in the laboratory and known to be carcinogenic in animals (rats, mice...).

ALKYLATING AGENT	MUTAGENICITY (AMES test)		CARCINOGENICITY IN RATS
	SIMPLE	MODIFIED (+ microsomes)	
Dimethyl sulfate	+		- *Subcutaneous* sarcoma + pulmonary metastasis - *Inhalation* Nasal cavity cancer
Methyl methanesulfonate	+-	+	- *Subcutaneous* sarcoma + cancer of the nervous system
Methyl iodide	-	+	- *Subcutaneous* sarcoma + pulmonary metastasis
Diazomethane	++		- *Inhalation* lung cancer

Table 1.14   Mutagenic and carcinogenic potencies of various alkylating agents.

in animals are potentially carcinogenic in human beings. They must thus be handled accordingly.

Table 1.14 summarizes the known data concerning the mutagenic and carcinogenic potency in animals of the principal alkylating agents commonly used in the laboratory.

The classification of carcinogenic substances

The classification of carcinogenic or potentially carcinogenic substances in humans thus constitutes an important element in the prevention of professional cancers. Various international organizations have contributed to this effort by establishing lists of carcinogenic compounds.

In the United States, the Gene-Tox Program of the Environmental Protection Agency (EPA) published, in 1987, data concerning 506 preselected chemical compounds. Four categories were defined which were then divided into subgroups. Depending on the results obtained, these 506 compounds were classed into the following four principal categories:

–Positive results (351 compounds)
–Negative results (61 compounds)
–Equivocal results (1 compound)
–Insufficient results (93 compounds)

Moreover, since 1969, the International Agency for Research on Cancer (IARC) in Lyon, founded by the World Health Organization (WHO), has undertaken the evaluation of the carcinogenic potencies of a variety of chemical substances and industrial processes. The IARC regularly publishes monographs (50 so far) based on the studies of commissions of international experts. In the seventh supplement of these monographs, which appeared in early 1988, data concerning 630 substances or industrial processes were summarized. These were classed into five categories:

–Group 1: the substance is *carcinogenic* in humans
–Group 2A: the substance is *probably carcinogenic* in humans
–Group 2B: the substance is *possibly carcinogenic* in humans
–Group 3: the carcinogenicity of the substance in humans *cannot be unambiguously determined*
–Group 4: the substance is *probably not carcinogenic* in humans

Group 1: Agents known to be carcinogenic in humans.   Agents in Group 1 include:

–12 industrial processes and situations of exposure
–pharmaceuticals (medication, contraceptive drugs, etc.)
–7 complex mixtures (soot, tars, mineral oils, fish oils, asbestos-containing talc, etc.)
–3 social habits (tobacco, betel, etc.)
–13 common chemical substances (Table 1.15)

Group 2A: Substances that are probably carcinogenic in humans.   In group 2A are found all the substances generally recognized to be carcinogenic in people and for which experimental evidence, especially long-term studies in animals, is fairly conclusive. These substances are summarized in Table 1.16.

Aflatoxins	Bis(chloromethyl) ether and
Asbestos	Chloromethyl methyl ether
4-Aminobiphenyl	(technical grade)
Arsenic and its compounds	Vinyl chloride
Benzene	Hexavalent chromium compounds
Benzidine	Sulfur mustards (yperite)
	2-Naphthylamine
	Nickel and its compounds

Table 1.15   Chemical substances carcinogenic in humans (the underlined compounds are commonly used in laboratories).

Acrylonitrile
Benz(a)anthracene, benzo(a)pyrene, dibenz(a,h)anthracene
Beryllium and its compounds
Vinyl bromide
Cadmium and its compounds
Dimethylcarbamoyl chloride
Benzidine-based dyes
Creosotes
1,2-Dibromoethane
Epichlorohydrin
Formaldehyde
4,4'-Methylene bis(2-chloroaniline) (MCCA)
Nitrogen mustards
N-Ethyl-N-nitrosourea (ENU), Methyl nitrosourea (MNU)
N-Methyl-3-nitro-N-nitrosoguanidine (MNNG)
N-Nitrosodiethylamine, N-Nitrosodimethylamine
Ethylene oxide, propylene oxide, styrene oxide
Polychlorinated biphenyls (PCBs)
Crystalline silicic acid
Dimethyl sulfate, diethyl sulfate
Tris(2,3-dibromopropyl)phosphate (TRIS-BP)*
+ 9 pharmaceuticals (medication, anabolizing androgenic steroids,
5-methoxypsoralen...)

Table 1.16 Chemical substances that are probably carcinogenic in humans (the underlined compounds are commonly used in laboratories).

*Tris (2,3-dibromopropyl)phosphate or TRIS-BP, carcinogenic, should not be confused with tris(hydroxymethyl)aminomethane or TRIS(TRIS-Amino), an amino alcohol used as a buffer in biology and which is practically nontoxic.

$$O = P \begin{array}{l} -\ O\text{-}CH_2\text{-}CHBr\text{-}CH_2Br \\ -\ O\text{-}CH_2\text{-}CHBr\text{-}CH_2Br \\ -\ O\text{-}CH_2\text{-}CHBr\text{-}CH_2Br \end{array} \qquad H_2N - \overset{CH_2OH}{\underset{CH_2OH}{|}} - CH_2OH$$

TRIS - BP   TRIS buffer

Group 2B: Substances considered to be possibly carcinogenic in humans. In group 2B are included substances for which long-term animal experiments give sufficient evidence of their carcinogenic properties but for which data in human beings provide only inadequate evidence of carcinogenicity. Table 1.17 summarizes the principal substances belonging to group 2B, classified by chemical family.

These lists can neither be considered exhaustive nor definitive, but are intended to be modified and completed as research by the IARC group progresses.

Hydrocarbons :	1,3-butadiene, styrene
Chloride derivatives :	methylene chloride, chloroform, carbon tetrachloride, 1,2-dichloroethane, benzyl chloride, benzylidene chloride, phenylchloroform, tetrachloro-(or perchloro-)ethylene, p-dichlorobenzene, hexachlorobenzene, gamma isomer of hexachlorocyclohexane (lindane...), DDT..., chlorophenoxy herbicides (2,4-D, 2,4,5-T)
Oxygen derivatives :	diepoxybutane, 1,4-dioxane, t-butylhydroxyanisole (BHA), safrole, dihydrosafrole, acetaldehyde, ethyl acrylate, di(2-ethylhexyl)phthalate, β-butyrolactone, β-propiolactone, chlorophenols, dioxin(TCDD)...
Nitrogen derivatives :	2-nitropropane, 2-methylaziridine, o-toluidine, o-anisidine, p-cresidine, o-tolidine, o-dianisidine, o-dichlorobenzidine, acetamide, acrylamide, phenobarbital, hydrazine, dimethylhydrazines, p-aminoazobenzene, p-dimethylaminoazobenzene, o-aminoazotoluene, urethane (ethyl carbamate), toluene diisocyanates, bleomycin, chloramphenicol..., metronidazole, niridazole...
Sulfur derivatives :	methyl and ethyl methanesulfonates, thioacetamide, thiourea, 1,3-propanesultone, methylthiouracil, propylthiouracil, saccharin...
Phosphorus derivatives :	hexamethylphosphoric triamide (HMPT)
Mineral derivatives :	potassium bromate, lead and inorganic lead compounds

etc.

Table 1.17 Substances considered to be possibly carcinogenic in humans (the underlined compounds are commonly used in laboratories).

Based on data collected by both the IARC in Lyon and the EPA in the United States, it is possible to establish the principal substances that are genotoxic (or suspected to be so) in human beings and which are routinely handled in research laboratories. In Table 1.18 are collected 105 substances or families of substances (e.g., PCB, metals, etc.) frequently encountered in laboratories that employ chemicals. Among these are included starting materials used in the chemical industry such as various monomers (1,3-butadiene, styrene, vinyl chloride, ethylene oxide, ethyl acrylate, acrylamide, acrylonitrile, etc.), laboratory reagents (methyl iodide, 2,4-dinitro-1-fluorobenzene, benzidine, Michler's ketone, 2,4-dinitrophenyl-hydrazine, diazomethane, trypan blue, ethidium bromide, dimethyl sulfate, trimethyl phosphate, hexavalent chromium salts, etc.) and many commonly used solvents (benzene, methylene chloride, chloroform, carbon tetrachloride, 1,2-dichloroethane, trichloroethylene, perchloroethylene, 1,4-dioxane, 2-nitropropane, HMPT, etc.).

In this table, by no means exhaustive, are indicated:

–the chemical family each substance belongs to
–the substance's common name and its structural formula
–the chemical abstracts (CAS) number
–the most common use for the substance (M = monomer; R = reagent; S = solvent)
–the mutagenic activity as determined by different mutagenicity tests in bacteria, in particular, by the Ames test

The Ames test consists of measuring the capacity of a given substance to modify the DNA of bacteria (*Salmonella typhimurium*) in the presence or not of an enzymatic system of metabolism (mammalian hepatic microsomes). Mutagenic substances that act by an oxidative mechanism (hydrogen peroxide, etc.) may be detected by a particular strain of *Salmonella* (TA 102 strain). Using these tests, it has been estimated that 80% of substances that are mutagenic in the Ames tests are also carcinogenic in animals.

Carcinogenic potency in laboratory animals. Carcinogenic potency in laboratory animals is determined after administration of different doses of a substance over a prolonged period in various species (rat, mouse, hamster, etc.). Carcinogenic potency can vary depending on the animal species used and present a more or less significant selectivity for a particular organ. A positive long-term carcinogenicity test in at least one species of animals (usually a rodent) is a good indication that the substance is potentially carcinogenic in human beings, though it is not possible to predict which organ will be affected. In the IARC classification, substances in group 2 include those that are probably carcinogenic in human beings (group 2A) and for which experimental evidence has been well established, especially in animal experiments, as well as substances which are possibly carcinogenic in human beings (group 2B) and for which animal experiments are somewhat less clear-cut and evidence less well established.

The results of epidemiological studies. The increase in the number of cancers among workers exposed over prolonged periods to certain chemical substances (or mixtures) provided in a few cases irrefutable evidence of their carcinogenic potency in human beings. This is true for 50 of the substances, industrial processes, or living habits (tobacco, etc.) retained in group 1 by the IARC. In this case, sufficient proof exists of a cause-and-effect relationship between human exposure to a substance or a group of substances and the appearance of a given type of cancer.

The promoter effect.   Promoters are carcinogens that do not act directly on DNA but intervene by an epigenetic mechanism that varies depending on the class of promoter. For example, phorbol esters interact with a specific membrane receptor (protein kinase C, an enzyme whose natural ligand is diacylglycerol liberated by membrane-bound inositol phospholipids). Substances having co-carcinogenic activity, such as phenol have also been classed with those exhibiting promoter activity (e.g., skin cancer initiated by polyaromatic hydrocarbons), though they are not real promoters.

Classifications established by the IARC and the EPA (Gene-Tox Program). A certain number of substances included in the list have not yet been classified, though this does not mean that they do not demonstrate recognized genotoxic activity. This, for example, is the case for quinoline, a carcinogen in animals, and for ethidium bromide, a DNA intercalator, undeniably possessing mutagenic properties.

Proposed substitutes.   Solutions for the substitution of carcinogenic substances by noncarcinogenic ones are proposed, especially in the case of solvents and for certain substances such as benzidine and asbestos.

Principal families of carcinogenic substances

Potentially carcinogenic chemical substances often encountered in the laboratory have DNA as a common target. A large majority of these molecules may be considered electrophilic agents acting either directly or under the form of reactive intermediates (or reactive metabolites). The electrophilic nature of a molecule (before or after metabolism) is not by itself sufficient to impart carcinogenic properties. Thus, numerous acylating agents (acetic anhydride, phthalic anhydride, etc.) are not carcinogenic despite their considerable electrophilic nature. Moreover, nonelectrophilic molecules may possess carcinogenic properties; this is the case of asbestos, the fibers of which contain a majority of nucleophilic sites at the surface. Though a variety of mechanisms may intervene in chemical carcinogenesis, the interaction between an electrophilic entity and the nucleophilic sites present on macromolecules of the cell nucleus (essentially DNA but also proteins and RNA) constitutes a satisfactory hypothesis in keeping with the current state of knowledge.
    Most of the compounds presented in Table 1.18 may be grouped into the following major chemical families:

    −Aromatic hydrocarbons
    −Halogenated derivatives: alkanes, alkenes, compounds having a bis-

chloroethyl moiety linked to a heteroatom (O, S, N)
- Three-atom heterocyclic compounds (epoxides, aziridines, episulfides)
- Nitro derivatives
- Aromatic amines, hydrazine and derivatives
- N-nitroso derivatives
- Alkyl sulfates and sulfonates
- Inorganic substances (asbestos, arsenic, metals, etc.)

It should be noted that certain compounds such as formaldehyde, dioxane, or hexamethylphosphoric triamide (HMPT), do not fit into these categories.

Aromatic hydrocarbons. Except for benzene, responsible for leukemias in human beings, the aromatic hydrocarbons that are carcinogenic in animals are polycyclic compounds: benzo(a)pyrene, dibenzo(a,h)anthracene, and so on.

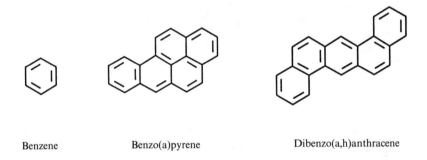

Benzene   Benzo(a)pyrene   Dibenzo(a,h)anthracene

These polycyclic hydrocarbons (polyarenes) may be formed during the course of the incomplete combustion of organic matter (laboratory fires, etc.). They are also found in tobacco smoke and may potentiate the carcinogenic action of other chemical compounds introduced into the organism via the respiratory system.

Halogenated derivatives

Haloalkanes. Because of their powerful electrophilic properties, many alkyl halides are mutagenic and some, carcinogenic in animals. In keeping with the order of their electrophilic strengths, the mutagenic potency of haloalkanes generally increases from chlorides to iodides. Alkylfluorides are practically inactive.

Halogenated methane derivatives present the most serious problems in laboratories. These mono- and dihalogenated derivatives (CH_3Cl, CH_3Br, CH_3I, CH_2Cl_2, etc.) are mutagenic and some, carcinogenic (CH_3I). Chloroform ($CHCl_3$) and carbon tetrachloride (CCl_4) are carcinogenic in animals (liver and kidney cancers) and for CCl_4, there have been several reports of hepatomas in people.

Within the possible series of methane homologues, the polyhalogenated derivatives are the most active. These include various solvents that are carcinogenic in animals: 1,2-dichloroethane, 1,1,2-trichloroethane, 1,1,2,2-tetrachloroethane, and so on. 1,1,1-Trichloroethane (Cl_3C—CH_3) does not appear to have this inconvenience.

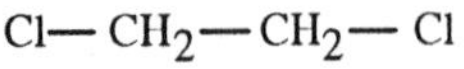

$$Cl—CH_2—CH_2—Cl$$

1,2-Dichloroethane

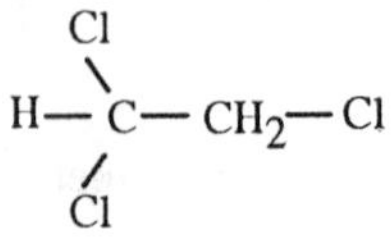

1,1,2-Trichloroethane

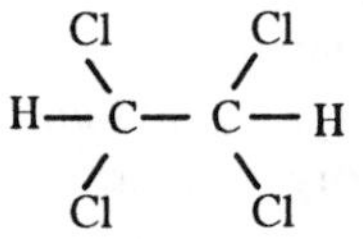

1,1,2,2-Tetrachloroethane

A particular case is that of derivatives having one or several benzylic halogens, of which several representatives have been shown to be mutagenic and carcinogenic in animals. Thus, benzyl chloride, benzal chloride, and phenylchloroform (α,α',α''-trichlorotoluene) are carcinogenic in animals and probably in people (lung cancer and maxillary lymphoma).

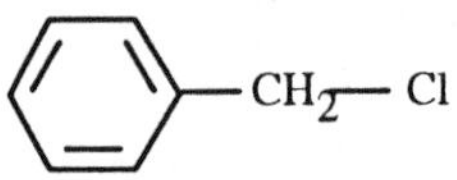

Benzyl chloride

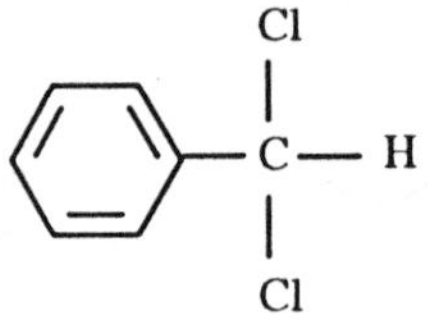

Benzal chloride

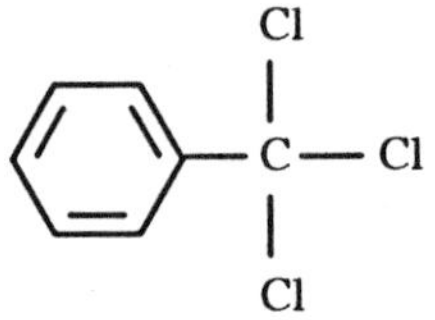

Phenylchloroform

Haloalkenes.   Most of the chlorinated derivatives of ethylene:

vinyl chloride (R_1=Cl, R_2=R_3=R_4=H)
vinylidene chloride (R_1=R_2=Cl, R_3=R_4=H)
trichloroethylene (R_1=R_2=R_3=Cl, R_4=H)
perchloroethylene (R_1=R_2=R_3=R_4=Cl)

are mutagenic and carcinogenic in animal experiments. Only vinyl chloride is a recognized carcinogen in people (angiosarcoma of the liver).

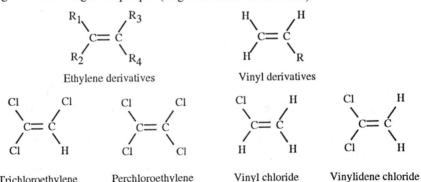

Ethylene derivatives

Vinyl derivatives

Trichloroethylene   Perchloroethylene   Vinyl chloride   Vinylidene chloride

Among the nonhalogenated derivatives of ethylene, styrene (R_1=C_6H_5, R_2=R_3=R_4=H) and acrylonitrile, (R_1=CN, R_2=R_3=R_4=H)) are also carcinogenic in animals. However, not all compounds having a vinyl moiety, H_2C=CHR, are necessarily carcinogenic.

Recently, acrylamide (R—$CONH_2$), a potent neurotoxin, has been shown to be carcinogenic in animals (classed in Table 2B by the IARC).

| Styrene | Acrylonitrile | Acrylamide |

α- and β-haloethers.   Many ethers possessing one or two chlorine atoms α or β to the oxygen function demonstrate mutagenic and carcinogenic properties in animals.

Bis(chloromethyl) ether (BCME), a potent carcinogen in human beings

α-monochloroether   α,α'-dichloroether

(bronchial cancer), can be formed together with chloromethyl methyl ether (CMME), when formaldehyde comes into contact with hydrochloric acid or sodium hypochlorite.

These two colorless, highly volatile liquids rapidly decompose in water

H_3C-O-CH_2-Cl $ClCH_2$-O-CH_2Cl

Chloromethyl methyl ether   Bis(chloromethyl) ether

releasing hydrochloric acid and formaldehyde and, in the case of chloromethyl methyl ether, methanol. These substances are widely used in laboratories and in industry (ion-exchange resins, insecticides, etc.). Chloromethyl methyl ether, is especially useful in stabilizing alcohols, phenols, and carboxylic acids and for the halomethylation of aromatic compounds.

Chloromethyl methyl ether is classically prepared by saturating a mixture of formol and methanol with hydrochloric acid gas (Figure 1.13).

Variable quantities of bis(chloromethyl) ether are also formed in this

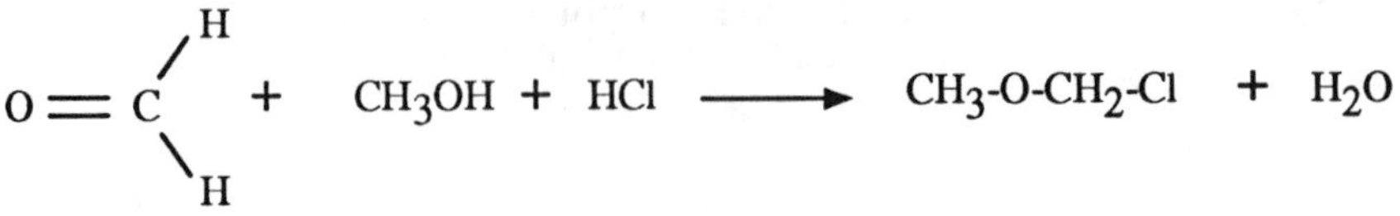

$$O = C \underset{H}{\overset{H}{\big\langle}} \quad + \quad CH_3OH \ + \ HCl \quad \longrightarrow \quad CH_3\text{-}O\text{-}CH_2\text{-}Cl \quad + \quad H_2O$$

Figure 1.13   Preparation of chloromethyl methyl ether.

reaction such that commercial solutions of chloromethyl methyl ether always contain 1–8% of this contaminant.

Bis(chloromethyl) ether is highly carcinogenic in a variety of animals (rats, mice, etc.) and, depending on the route of administration, skin, nasal cavity, or lung tumors are produced. By comparison, chloromethyl methyl ether is suspected to be a weaker carcinogen, but since this substance is always contaminated with BCME, it is in fact difficult to determine its intrinsic activity. Moreover, BCME is formed whenever formaldehyde and hydrochloric acid fumes come into contact. These mixtures, routinely used in industry (textiles, etc.) and for disinfecting rooms (sterilization rooms) or surgical equipment, are very toxic. A survey of workers in contact with α-haloethers (CMME and BCME) has revealed a higher risk of bronchial cancers in these subjects.

Certain β-haloethers are also carcinogenic in animals; a few are used in

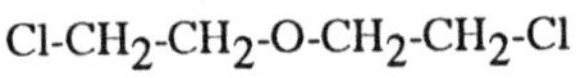

$$Cl\text{-}CH_2\text{-}CH_2\text{-}O\text{-}CH_2\text{-}CH_2\text{-}Cl \qquad\qquad H_2C\text{-}CH\text{-}CH_2Cl$$
$$\underset{O}{\diagdown\diagup}$$

Bis(2-chloroethyl) ether 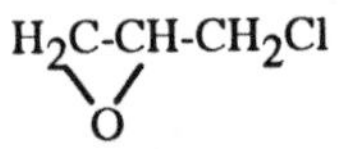   Epichlorohydrin

synthesis: bis(2-chloroethyl) ether (BCE), and epichlorohydrin, which incorporates an epoxide in its structure.

Epichlorohydrin produces skin cancer in mice and may be responsible for bronchial cancers in workers subject to long-term exposures (15 years) to this substance.

Compounds having a bis(chloroethyl) moiety linked by a heteroatom.   Compounds having two chloroethyl chains ($-CH_2=CH_2Cl$) joined by a heteroatom (S, N, and O) are mutagenic and carcinogenic. Among the mustard derivatives (Z$=$S or N$-$R), potent carcinogens in humans (lung cancer), yperite (Z$=$S), is used in organic synthesis.

$$Z \overset{CH_2\text{-}CH_2\text{-}Cl}{\underset{CH_2\text{-}CH_2\text{-}Cl}{<}}$$

Bis(chloroethyl) derivatives
Z = S or N - R

Three-atom heterocycles.   Compounds having three-atom rings, one of which is a heteroatom (O, N, S), possess ring strain which increases their electrophilic character. As a result, many of these are mutagenic and carcinogenic in animals.

$$Z = \begin{cases} O & \text{Epoxides} \\ \left.\begin{matrix}N\text{-}H \\ N\text{-}R\end{matrix}\right\} & \text{Aziridines} \\ S & \text{Episulfides} \end{cases}$$

Many simple epoxides (Z=O) commonly used in the laboratory (ethylene oxide, propylene oxide, etc.) are mutagenic and carcinogenic in animals and, moreover, cases of leukemia have been reported in workers exposed to ethylene oxide. Chlorinated derivatives of ethylene are metabolized into epoxides which could be responsible for the carcinogenic activity of certain ethylene compounds. Numerous aziridines (Z=N—H or N—R) used in synthesis, such as aziridine or 2-methylaziridine, are mutagenic and carcinogenic. The same is true of ethylene sulfide (Z=S).

It should be noted that, among four-atom cyclic compounds possessing a heteroatom, β-propiolactone (and its methylated derivatives) has been shown

| Ethylene oxide | Propylene oxide | Aziridine | 2-Methyl-aziridine | Ethylene sulfide |

to be a potent carcinogen in animals.

β-propiolactone

Table 1.18 Principal genotoxic substances commonly used in laboratories (mutagenic-carcinogenic substances).

N°	Chemical Family	Chemical Substance Common Name Structure	CAS N°	Use	Mutagenicity (Ames Test)	Carcinogenicity Animal	Carcinogenicity Epidemio. Study	Promoter Effect	Classification IARC	Classification Gene-Tox	Suggested Replacement Compounds
1	Hydrocarbons	H_2C⟍⟍CH_2 transoid 1,3-Butadiene H_2C⟍⟍CH_2 cisoid	106-99-0	M	+	+			2B	+	
2		Benzene	71-43-2	S	-	+	+		1	+	Cyclohexane Toluene
3		Styrene ⟍CH_2	100-42-5	M	(+)	(+)			2B	(+)	
4		Benzo(a)pyrene B(a)P	50-32-8	R	+	+			2A	+	

N°	Chemical Family	Chemical Substance Common Name Structure	CAS N°	Use	Mutagenicity (Ames Test)	Carcinogenicity		Promoter Effect	Classification		Suggested Replacement Compounds
						Animal	Epidemio. Study		IARC	Gene-Tox	
5	Hydrocarbons	7,12-Dimethyl-benz(a)anthracene (DMBA)	57-97-6	R	+	+				+	
6		3-Methylcholanthrene (3-MC)	56-49-5	R	+	+				+	
7	Halogenated derivatives	Methyl iodide	74-88-4	R	+	(+)			3	+	
8	- Haloalkanes	Methylene chloride	75-09-2	S	+	+			2B	+	1,1,1-Trichloroethane

N°	Chemical Family	Chemical Substance Common Name Structure	CAS N°	Use	Mutagenicity (Ames Test)	Carcinogenicity Animal	Epidemio. Study	Promoter Effect	Classification IARC	Gene-Tox	Suggested Replacement Compounds
9		Chloroform (Trichloromethane)	67-66-3	S	-	+			2B	+	1,1,1-Trichloroethane
10		Carbon tetrachloride	56-23-5	S	-	+			2B	+	1,1,1-Trichloroethane
11		1,2-Dichloroethane	107-06-2	S	+	+			2B	+	1,1,1-Trichloroethane
12		Benzyl chloride	100-44-7	R	+	(+)			2B	(+)	
13		Phenylchloroform (α,α',α"-Trichlorotoluene)	98-07-7	R	+	+			2B	+	

N°	Chemical Family	Chemical Substance Common Name Structure	CAS N°	Use	Mutagenicity (Ames Test)	Carcinogenicity Animal	Epidemio. Study	Promoter Effect	Classification IARC	Gene-Tox	Suggested Replacement Compounds
14		DDT (4,4'-Dichlorodiphenyl-trichloroethane)	50-29-3	R	-	+		+	2B	+	
15	- Chloroalkenes	Vinyl chloride	75-01-4	M	+	+	+		1	+	1,1,1-Trichloro-ethane
16		Trichloroethylene (trichlo)	79-01-4	S	+	(+)			3	(+)	1,1,1-Trichloro-ethane
17		Perchloroethylene (perchlo)	127-18-4	S	+	+			2B	(+)	1,1,1-Trichloro-ethane

N°	Chemical Family	Chemical Substance Common Name Structure	CAS N°	Use	Mutagenicity (Ames Test)	Carcinogenicity Animal	Carcinogenicity Epidemio. Study	Promoter Effect	Classification IARC	Classification Gene-Tox	Suggested Replacement Compounds
18	- Chloroarenes	Polychlorobiphenyls (P.C.B.)	1336-36-3	R	-	+	(+)	+	2A		Silicones
19	Oxygen Derivatives - Phenols	Phenol —OH	108-95-2	R	-	-		+			
20		2,4,6-Trichlorophenol	88-06-2	R	-	+			2B	+	
21	- Ethers	Ethylene oxide	75-21-8	R M	+	+	(+)		2A	+	
22		Propylene oxide	75-56-9	R M	+	+			2A	+	
23		Chloromethyl methyl ether (CMME) (Technical grade) $CH_3 - O - CH_2 - Cl$	107-30-2	R	+	+	+		1	(+)	

N°	Chemical Family	Chemical Substance Common Name Structure	CAS N°	Use	Mutagenicity (Ames Test)	Carcinogenicity		Promoter Effect	Classification		Suggested Replacement Compounds
						Animal	Epidemio. Study		IARC	Gene-Tox	
24		Bis(chloromethyl) ether (BCME) Cl - CH$_2$ - O - CH$_2$ - Cl	542-88-1	R	+	+	+		1	+	
25		Epichlorohydrin CH$_2$Cl	106-89-8	R	+	+			2A	+	
26		1,4-Dioxane	123-91-1	S	-	+			2B	+	tetrahydrofuran
27		Dioxin (2,3,7,8-tetrachloro-dibenzo-p-dioxin)	1746-01-6	R	-	+		+	2B	+	
28	- Aldehydes	Formaldehyde	50-00-0	R	-	+		+	2B	+	

N°	Chemical Family	Chemical Substance Common Name Structure	CAS N°	Use	Mutagenicity (Ames Test)	Carcinogenicity Animal	Epidemio. Study	Promoter Effect	Classification ICRC	Gene-Tox	Suggested Replacement Compounds
29		Acetaldehyde	75-07-0	R	+	+			2B		
30		Glyoxal	107-22-2	R	+						
31	- Esters	Ethyl acrylate	140-88-5	M	-	+			2B		
32		Di(2-ethylhexyl)phthalate	117-81-7	R	-	+		+	2B	+	
33		Ethyl pyrocarbonate	1609-47-8	R	-	+					

N°	Chemical Family	Chemical Substance Common Name Structure	CAS N°	Use	Mutagenicity (Ames Test)	Carcinogenicity Animal	Epidemio. Study	Promoter Effect	Classification IARC	Gene-Tox	Suggested Replacement Compounds
34		Phorbol myristate acetae $CH_3(CH_2)_{12}COO$ $OCOCH_3$ H_3C ... H, H_3C ..., OH, HO, O, CH_2OH	16561-29-8	R	-	+		+			
35	- Lactones	β-propiolactone	57-57-8	R	+	+			2B	+	
36		β-butyrolactone	3068-88-0	R	+	+			2B	+	
37	- Other oxygen derivatives	Aflatoxin B$_1$	1162-65-8	R	+	+	+		1	+	

N°	Chemical Family	Chemical Substance Common Name Structure	CAS N°	Use	Mutagenicity (Ames Test)	Carcinogenicity Animal	Epidemio. Study	Promoter Effect	Classification IARC	Gene-Tox	Suggested Replacement Compounds
38	Nitrogen Derivatives - Nitroalkanes	H_3C $\overset{H}{\underset{NO_2}{\underset{\textstyle}{X}}}$ H_3C 2-Nitropropane	79-46-9	S	+	+			2B	+	1-Nitropropane $C_3H_7NO_2$
39	- Nitroarenes	4-Nitrobiphenyl NO_2	92-93-3	R	+	(+)			3	+	
40		CH_3 NO_2 NO_2 2,4-Dinitrotoluene	121-14-2	R	+	+					
41		1-Fluoro-2,4-dinitrobenzene (2,4-DNFB) F NO_2 NO_2	70-34-8	R	+			+			

N°	Chemical Family	Chemical Substance Common Name Structure	CAS N°	Use	Mutagenicity (Ames Test)	Carcinogenicity Animal	Carcinogenicity Epidemio. Study	Promoter Effect	Classification IARC	Classification Gene-Tox	Suggested Replacement Compounds
42	- Amines	Aziridine (Ethylenimine)	151-56-4	R	+	(+)			3	(+)	
43		Methyl aziridine (Propylenimine)	75-55-8	R	+	+			2B	+	
44		Nitrilotriacetic acid (NTA) and its sodium salt	139-13-9 and 5064-31-3	R	-	+				+	
45		Aniline	62-53-3	S R	-	(+)			3		

N°	Chemical Family	Chemical Substance Common Name Structure	CAS N°	Use	Mutagenicity (Ames Test)	Carcinogenicity Animal	Carcinogenicity Epidemio. Study	Promoter Effect	Classification IARC	Classification Gene-Tox	Suggested Replacement Compounds
46		o-Toluidine (NH₂, CH₃)	95-53-4	S R	-	(+)			2B	+	
47		o-Anisidine (NH₂, OCH₃)	90-04-0	R	+	+			2B	+	
48		2-Naphthylamine (β-Naphthylamine) (NH₂)	91-59-8	R	+	+	+		1	+	
49		4-Aminobiphenyl (NH₂)	92-67-1	R	+	+	+		1	+	
50		Benzidine (H₂N, NH₂)	92-87-5	R	+	+	+		1	+	3,3',5,5'-Tetramethyl-benzidine

N°	Chemical Family	Chemical Substance Common Name Structure	CAS N°	Use	Mutagenicity (Ames Test)	Carcinogenicity Animal / Epidemio. Study	Promoter Effect	Classification IARC	Gene-Tox	Suggested Replacement Compounds
51		o-Tolidine (3,3'-dimethylbenzidine)	119-93-7	R	+	+		2B	+	
52		o-Dianisidine (3,3'-dimethoxybenzidine)	119-90-4	R	+	+		2B	+	
53		o-Dichlorobenzidine (3,3'-dichlorobenzidine)	91-94-1	R	+	+		2B	+	
54		o-Diaminobenzidine (3,3'-diaminobenzidine)	91-95-2	R	+	(+)				

N°	Chemical Family	Chemical Substance Common Name Structure	CAS N°	Use	Mutagenicity (Ames Test)	Carcinogenicity Animal	Epidemio. Study	Promoter Effect	IARC	Gene-Tox	Suggested Replacement Compounds
55		Michler's ketone Bis(4,4'-dimethylamino) benzophenone	90-94-8	R	+	+				+	
56	- Amides	Acetamide	60-35-5	R	-	+			2B		
57		Acrylamide	79-06-1	M	-	+			2B		
58		2-Acetylaminofluorene (2-AAF)	53-96-3	R	+	+				+	

N°	Chemical Family	Chemical Substance Common Name Structure	CAS N°	Use	Mutagenicity (Ames Test)	Carcinogenicity Animal	Carcinogenicity Epidemio. Study	Promoter Effect	Classification IARC	Classification Gene-Tox	Suggested Replacement Compounds
59	- Ureides	Phenobarbital	50-06-6	R	-	+		+	2B		
60	- Nitriles	Acrylonitrile $H_2C{=}{\cdot}N$	107-13-1	M	+	+	(+)		2A	+	
61	- Hydrazines	Hydrazine H_2N-NH_2 / Hydrazine hydrate H_2N-NH_2, H_2O	302-01-2 and 7803-57-8	R	+	+			2B	+	
62		1,2-Dimethylhydrazine	540-73-8	R	+	+			2B	+	

N°	Chemical Family	Chemical Substance Common Name Structure	CAS N°	Use	Mutagenicity (Ames Test)	Carcinogenicity Animal Epidemio. Study		Promoter Effect	Classification IARC Gene-Tox		Suggested Replacement Compounds
63		1,1-Dimethylhydrazine $H_3C-N-NH_2$ (H_3C)	51-14-7	R	+	+			2B	+	
64		Phenylhydrazine $HN-NH_2$	100-63-0	R	-	+					
65		2,4-Dinitrophenylhydrazine O_2N / NO_2 / $HN-NH_2$	119-26-6	R	+	(+)					
66		Hydrazobenzene (1,2-Diphenylhydrazine) $HN-NH$	122-66-7	R	+	+				+	

N°	Chemical Family	Chemical Substance Common Name Structure	CAS N°	Use	Mutagenicity (Ames Test)	Carcinogenicity Animal	Epidemio. Study	Promoter Effect	Classification IARC	Gene-Tox	Suggested Replacement Compounds
67	- Nitrosoamines	N-Nitrosodimethylamine (NDMA) H3C—N—N=O / H3C	62-75-9	S R	+	+			2A	+	
68		N-Nitrosodiethylamine H3CH2C—N—N=O / H3CH2C	55-18-5	R	+	+			2A	+	
69	- Nitrosoureas	N-Nitroso-N-methylurea H3C—N—N=O / H2N—C=O	684-93-5	R	+	+			2A	+	
70	- Nitroso-guanidines	N-Methyl-N'-nitro-N-nitrosoguanidine (MNNG) H3C—N—N=O / O2N—N—NH / H	70-25-7	R	+	+			2A	+	

N°	Chemical Family	Chemical Substance Common Name Structure	CAS N°	Use	Mutagenicity (Ames Test)	Carcinogenicity Animal	Carcinogenicity Epidemio. Study	Promoter Effect	Classification IARC	Classification Gene-Tox	Suggested Replacement Compounds
71	- Nitroso-hydroxylamines	Cupferron (ammonium salt of N-nitrosophenylhydroxylamine)	135-20-6	R	+	+				+	
72	- Diazo compounds	Diazomethane $H_2C^- - N^+ \equiv N$	334-88-3	R	+	+			3	+	
73	- Azo compounds	Azobenzene	103-33-3	R	+	+			3	+	
74		Trypan blue (TPB)	72-57-1	R	+	+			2B	+	

N°	Chemical Family	Chemical Substance Common Name Structure	CAS N°	Use	Mutagenicity (Ames Test)	Carcinogenicity Animal	Carcinogenicity Epidemio. Study	Promoter Effect	Classification IARC	Classification Gene-Tox	Suggested Replacement Compounds
75	- Isocyanates	Toluyldiisocyanate (TDI) Mixture of 2,4 and 2,6 isomers	91-08-7 (2,6) / 584-84-9 (2,4)	M	+	+			2B		
76	- Carbamates	Urethane (ethyl carbamate)	51-79-6	R	+	+			2B	+	
77	- Heterocyclic bases	3-Amino-9-ethylcarbazole (HCl)	132-32-1	R	+	+				+	

N°	Chemical Family	Chemical Substance Common Name Structure	CAS N°	Use	Mutagenicity (Ames Test)	Carcinogenicity Animal	Carcinogenicity Epidemio. Study	Promoter Effect	Classification ICRC	Classification Gene-Tox	Suggested Replacement Compounds
78		Quinoline	91-22-5	S / R	+	+					
79		4-Nitroquinoline N-oxide	56-57-5	R	+	+				+	
80		Ethidium bromide	1239-45-8	R	+						Propidium iodide

N°	Chemical Family	Chemical Substance Common Name Structure	CAS N°	Use	Mutagenicity (Ames Test)	Carcinogenicity Animal	Epidemio. Study	Promoter Effect	IARC	Gene-Tox	Suggested Replacement Compounds
81		Mitomycin C	50-07-7	R	+	+			2B	+	
82	Sulfur Derivatives	Yperite (2-chloroethylsulfide)	505-60-2	R		(+)	+		1	(+)	
83		Thioacetamide	62-55-5	R	-	+			2B	+	
84		Thiourea	62-56-6	R	-	+			2B	+	

N°	Chemical Family	Chemical Substance Common Name Structure	CAS N°	Use	Mutagenicity (Ames Test)	Carcinogenicity Animal	Carcinogenicity Epidemio. Study	Promoter Effect	Classification IARC	Classification Gene-Tox	Suggested Replacement Compounds
85		Dimethyl sulfate $O=\!\!\!\!\overset{O-CH_3}{\underset{O-CH_3}{S}}$	77-78-1	R	+	+	(+)		2A	+	
86		Diethyl sulfate $O=\!\!\!\!\overset{O-C_2H_5}{\underset{O-C_2H_5}{S}}$	64-67-5	R		(+)	+		1	(+)	
87		Methyl methanesulfonate (MMS) $O=\!\!\!\!\overset{CH_3}{\underset{O-CH_3}{S}}$	66-27-3	R	+	+			2B	+	
88		Ethyl methanesulfonate (EMS) $O=\!\!\!\!\overset{CH_3}{\underset{O-C_2H_5}{S}}$	62-50-0	R	+	+			2B	+	
89		1,3-Propanesultone	1120-71-4	R	+	+			2B	+	

N°	Chemical Family	Chemical Substance Common Name Structure	CAS N°	Use	Mutagenicity (Ames Test)	Carcinogenicity Animal	Carcinogenicity Epidemio. Study	Promoter Effect	Classification IARC	Classification Gene-Tox	Suggested Replacement Compounds
90		d-Ethionine H_5C_2–S–…NH_2 COOH	67-21-0	R	–	+				+	
91	Phosphorus Derivatives	Trimethylphosphate $O=P(OCH_3)_3$	512-56-1	S / R	(+)	(+)	+		1	(+)	
92		Hexamethylphosphoric triamide (HMPT) $O^- – P^+(N(CH_3)_2)_3$	680-31-9	S	+	+			2B	+	$H_3C\,_N$ C=O $N\,CH_3$ N,N'-Dimethyl propyleneurea (DMPU)
93		Cyclophosphamide	50-18-0	R	+	+	+		1	+	

N°	Chemical Family	Chemical Substance Common Name Structure	CAS N°	Use	Mutagenicity (Ames Test)	Carcinogenicity Animal	Carcinogenicity Epidemio. Study	Promoter Effect	Classification IARC	Classification Gene-Tox	Suggested Replacement Compounds
94	Inorganic Compounds	Asbestos	1332-21-4	R		(+)	+	+	1	+	Ceramic fibers
95		Arsenic and certain derivatives			-	(+)	+		1		
		Arsine AsH$_3$	7784-42-1	R		(+)					
96		Sodium azide $N\equiv N^+ = N^-\ Na^+$	26628-22-8	R	+	-					
97		Beryllium and certain derivatives				+	(+)		2A		
		Beryllium	7440-41-7	R	-	+				+	
		Beryllium oxide	1304-56-9		-	+				+	
		Beryllium sulfate	13510-49-1		-	+				+	
98		Cadmium and certain derivatives			-	+	(+)		2A		
		Cadmium	7440-43-9	R		+				(+)	
		Cadmium chloride	10108-64-2			+				+	
		Cadmium oxide	1306-19-0			+				+	
		Cadmium sulfate	10124-36-4			+				+	
		Cadmium sulfide	1306-23-6			+				+	

N°	Chemical Family	Chemical Substance Common Name / Structure	CAS N°	Use	Mutagenicity (Ames Test)	Carcinogenicity — Animal	Carcinogenicity — Epidemio. Study	Promoter Effect	Classification — IARC	Classification — Gene-Tox	Suggested Replacement Compounds
99		Chromium and certain derivatives									
		Hexavalent compounds Alkaline chromates and dichromates								+	
		Chromium trioxide	1333-82-0	R					3	+	
		Trivalent compounds									
100		H$_2$N–Pt(Cl)–... H$_2$N ... Cl Cis-platin	15663-27-1	R	+	+			2A	+	
101		Hydrogen peroxide H - O - O - H and certain derivatives (benzoyl peroxide...)	7722-84-1					+			
		(benzoyl peroxide structure)	94-36-0	R	(+)	(+)					

N°	Chemical Family	Chemical Substance Common Name Structure	CAS N°	Use	Mutagenicity (Ames Test)	Carcinogenicity Animal	Carcinogenicity Epidemio. Study	Promoter Effect	Classification IARC Gene-Tox	Suggested Replacement Compounds
102		Nickel and certain derivatives		R	-	+				
		Nickel subsulfide Ni_2S_3	12035-72-2	R	-	+			+	
103		Lead and certain derivatives		R	-	+			2B	
		Lead phosphate	7446-27-7	R	-	+			+	
		Lead subacetate	1335-32-6		-	+			+	
		Lead (II) acetate	1301-04-2		-	+			+	
		Lead (IV) acetate	546-67-8			+			+	
104		Crystalline silica $(SiO_2)_n$	7631-86-9	R		+	(+)		2A	
105		Selenium sulfide SeS	7446-34-6	R		+			+	

Source: Taken from A. Picot and M. Castegnaro. *Risks associated with handling of carcinogenic substances.* Actualité Chimique, French Chemical Society, Paris (1989), pp. 12–27.

Nitro derivatives. Among the nitroalkanes, only 2-nitropropane is mutagenic and hepatocarcinogenic in rats, however, many aromatic nitro compounds are mutagenic and carcinogenic in animals, and this regardless of whether they possess a single ring (2,4-dinitrotoluene, etc.) or several (1-nitropyrene, 2,4,7-trinitro-fluorenone, etc.).

2-Nitropropane   2,4-Dinitrotoluene   1-Nitropyrene   2,4,7-Trinitrofluorenone

Nitroarenes are metabolized (by the liver and by intestinal flora) into nitrosoarenes and then into N-hydroxylated derivatives. These hydroxyl-amines, which are also formed by oxidative metabolism of aromatic amines, are considered to be the reactive intermediates (alone or in the form of an active ester) involved in the activity of the various nitrogenated aromatics. All these compounds must be considered potential carcinogens. Thus, 4-nitrobiphenyl, used in the synthesis of 4-aminobiphenyl (bladder carcinogen) is also a suspected carcinogen.

4-Nitrobiphenyl   4-Aminobiphenyl

In the 2-naphthylamine series, 2-naphthylhydroxylamine, 2-nitrosonaphthalene, and 2-nitronaphthalene are carcinogenic in animals.

2-Naphthylamine   2-Naphthylhydroxylamine

2-Nitrosonaphthalene   2-Nitronaphthalene

Heterocycles having a nitro functionality (2-nitrofurans, 5-nitroimidazoles, etc.) may possess mutagenic and carcinogenic properties.

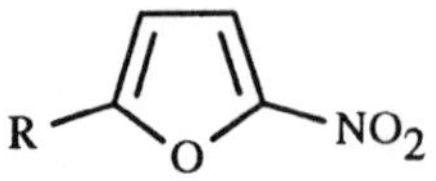

Substituted 2-nitrofurans

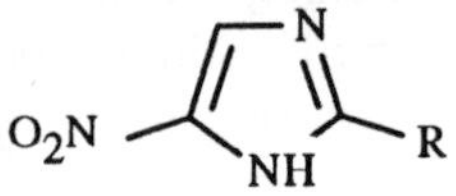

Substituted 5-nitroimidazoles

Aromatic amines.   It is among aromatic amines that are found the most potent carcinogens likely to be manipulated in the laboratory. In particular, 2-naphthylamine and benzidine (and the latter's chloro, methyl, methoxy, and amino derivatives) lead, after a relatively short latency period (5 years or less) to bladder cancers in animals and, in the case of benzidine, also in humans.   Certain derivatives of aniline (*ortho*-toluidine, and probably aniline itself) are also considered to be carcinogenic.

Among nitrogenated heterocyclic bases derived from aromatic amines, quinoline (and some of its derivatives) is carcinogenic in animals while isoquinoline is not.

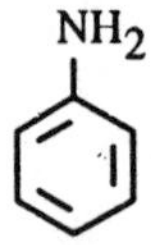

Aniline

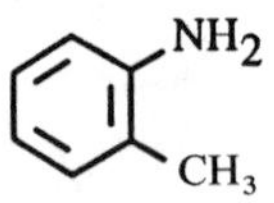

o-Toluidine

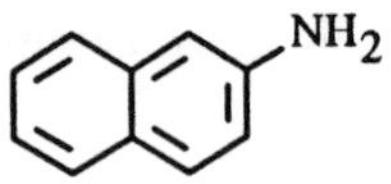

2-Naphthylamine

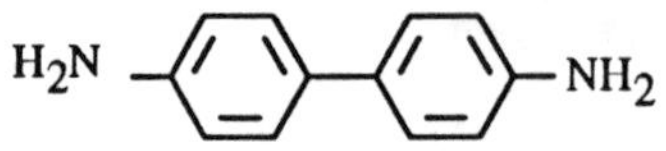

Benzidine

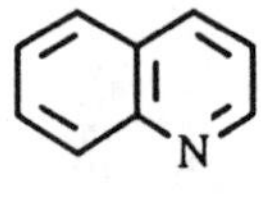

Quinoline

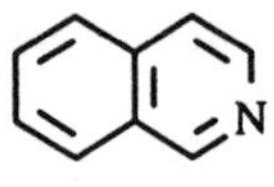

Isoquinoline

Hydrazine and its derivatives.   Numerous hydrazines, routinely used in the laboratory (hydrazine hydrate, various methylhydrazines, phenylhydrazine, etc.) are mutagenic and carcinogenic in animals. Hydrazines are suspected of causing cancer of the lungs, nervous system, liver, kidneys, hematopoietic organs, breast, and conjunctive tissue in humans.

Hydrazine hydrate   1,1-Dimethylhydrazine 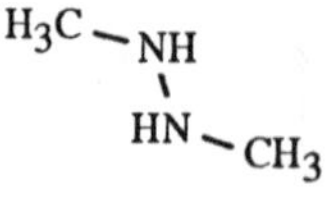   1,2-Dimethylhydrazine 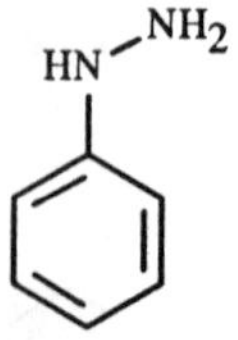   Phenylhydrazine

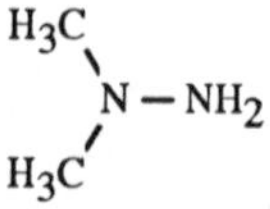

Certain hydrazine derivatives, such as the hydrazides (maleic hydrazide, semicarbazide, etc.) and azo derivatives (azobenzene, azo dyes, etc.) are carcinogenic in animals.

| Maleic hydrazide | Semicarbazide | Azobenzene |

N-Nitroso derivatives. Nitrosoamines, nitrosoureas, and nitrosoguanidines, recognized as carcinogens in animals, may also intervene in human carcinogenesis.

| Nitrosoamines | Nitrosoureas | Nitrosoguanidines |

Trace quantities of nitrosoamines are sometimes found in simple amines.

Nitrosoureas are sometimes used to prepare diazoalkanes such as diazomethane, itself carcinogenic in rodents. Among the diazoalkanes used in laboratories, ethyl diazoacetate has been shown to be carcinogenic in animals.

| Diazomethane | Ethyl diazoacetate |

Diazomethane. Diazomethane, an explosive and flammable yellow gas, is generally prepared in situ (ether, etc.). These concentrated solutions are very unstable and may explode violently in the presence of rough surfaces (ground-glass joints), of trace solid impurities, or as a result of exposure to intense light. It may be obtained by treatment of various nitrosoamides (N-methylnitrosourea, N-methyl-N'nitro-N-nitrosoguanidine) in alkaline media.

$$H_2C^- \!\!-\!\! N^+ \!\!\equiv\!\! N \leftrightarrow H_2C \!\!=\!\! N^+ \!\!=\!\! N^-$$

These nitrosoamides are potent carcinogens which are relatively unstable. It is therefore preferable to use less toxic and more stable nitroso derivatives such as N-methyl-N-nitroso-p-toluenesulfonamide (Diazald, Aldrich Chemical Co.).

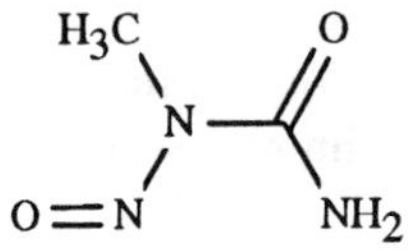

N-Methylnitrosourea

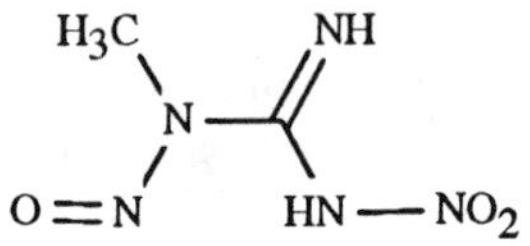

N-Methyl-N'-nitro-N-nitroso-guanidine

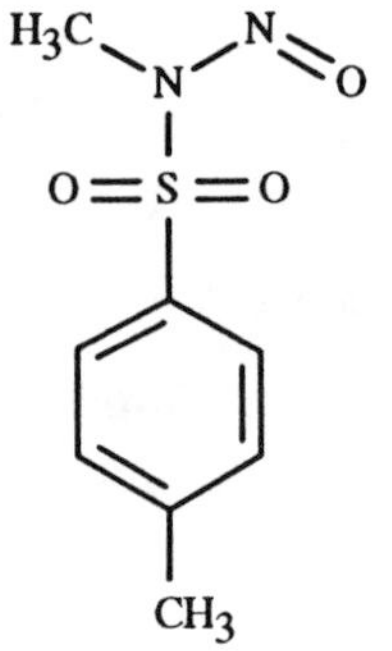

N-Methyl-N-nitroso-p-toluene-sulfonamide

Diazomethane, which can be used in neutral media, is an excellent methylating agent, especially for acidic compounds (acids, phenols, enols, etc.). It is also useful in a variety of syntheses (homologation of cyclic ketones, etc.). It is, however, an extremely toxic substance, strongly irritating to skin (dermatitis), ocular mucosa (conjunctivitis, corneal ulceration), and respiratory mucosa (dyspnoea, bronchitis, acute pulmonary edema). It may also provoke allergic reactions (asthma, etc.). Diazomethane also leads to pulmonary cancers in rats and mice (Table 1.18).

$$(CH_3)_3Si \!\!-\!\! CH \!\!=\!\! N^+ \!\!=\!\! N^-$$

Trimethylsilyldiazomethane

For the methylation of phenols, enols, and carboxylic acids as well as for homologation reactions of carbonyl compounds, diazomethane may be replaced by trimethylsilyldiazomethane, a stable reagent whose genotoxic activity is, however, not yet known.

Alkyl sulfates and sulfonates.   Alkyl sulfates and sulfonates are powerful electrophilic reagents and many of these derivatives are mutagenic and carcinogenic in animals.

Dimethyl sulfate.   Highly useful for the methylation of phenols, amines, and so on, dimethyl sulfate, was utilized experimentally as a toxic gas (Rationite) during World War I. It is an extremely dangerous substance whose hydrolysis

$$R_1 - O - \overset{\overset{\displaystyle O}{\|}}{\underset{\underset{\displaystyle O}{\|}}{S}} - O - R_2 \qquad H_3C - \overset{\overset{\displaystyle O}{\|}}{\underset{\underset{\displaystyle O}{\|}}{S}} - O - R \qquad F - \overset{\overset{\displaystyle O}{\|}}{\underset{\underset{\displaystyle O}{\|}}{S}} - O - R$$

Dialkyl sulfate   Alkyl methanesulfonate   Alkyl fluorosulfonate

within the organism releases sulfuric acid and methanol. In rats, mice, and hamsters, dimethyl sulfate, either inhaled or injected subcutaneously, produces localized cancers (Table 1.18). Moreover, administered to pregnant rats, it leads to multiple tumors (brain, liver, thyroid) in progeny. Many cases of bronchial cancer have also been observed among workers exposed to this industrial methylating agent.

Within the alkyl sulfate series, diethyl sulfate, is carcinogenic in rats while dibutyl sulfate, is inactive. Diisopropyl sulfate, may be the agent responsible for cancers of the nasal cavity and the larynx in employees of the isopropyl alcohol industry.

$$H_3C - O - \overset{\overset{\displaystyle O}{\|}}{\underset{\underset{\displaystyle O}{\|}}{S}} - O - CH_3 \qquad H_5C_2 - O - \overset{\overset{\displaystyle O}{\|}}{\underset{\underset{\displaystyle O}{\|}}{S}} - O - C_2H_5$$

Dimethyl sulfate   Diethyl sulfate

Diisopropyl sulfate   Dibutyl sulfate

Within this series of compounds, the methanesulfonates (methyl, ethyl) and methyl fluorosulfonate or "magic methyl" are powerful electrophilic agents that are strongly mutagenic and (for most) carcinogenic in animals.

Methyl methanesulfonate and ethyl methanesulfonate

These two colorless liquids are corrosive to skin and mucosa. Their ingestion provokes serious lesions in the digestive tract while their inhalation

$$H_3C \longrightarrow \overset{\overset{O}{\|}}{\underset{\underset{O}{\|}}{S}} - O - CH_3 \qquad\qquad H_3C \longrightarrow \overset{\overset{O}{\|}}{\underset{\underset{O}{\|}}{S}} - O - C_2H_5$$

Methyl methanesulfonate   Ethyl methanesulfonate

leads to strong bronchial irritation sometimes accompanied by congestion followed by acute pulmonary edema. The administration of a single dose of these substances in rodents is sufficient to induce tumor formation (Table 1.18).

Together with N-methylnitrosourea (MNU) and N-methyl-N'-nitro-N-nitrosoguanidine (MNNG), methyl methanesulfonate (MMS), and ethyl methanesulfonate (EMS), are routinely used in biology laboratories to produce bacterial mutations or to modify DNA. All these substances, strongly mutagenic and carcinogenic in rodents, must be handled with the utmost precaution.

Methyl fluorosulfonate.   Methyl fluorosulfonate, or "magic methyl" is a liquid of low volatility (bp = 92–94°C) which is more reactive than dimethyl sulfate. It is a potent methylating agent for amines, amides, nitriles, sulfur derivatives, and so on.

Methyl fluorosulfonate is an extremely dangerous reagent. In the vapor phase, its LC_{50} (lethal concentration in air with respect to 50% of animals exposed during a fixed period) in animals is of the order of 5–6 ppm. In 1976, a Dutch chemist died of pulmonary edema (6 h latency) after having been accidentally exposed to vapors resulting from the breaking of a 25 ml flask containing a solution of this reagent in methylene chloride.

Magic methyl apparently reacts without prior activation since it gives a positive Ames test for mutagenicity in the absence of hepatic microsomes. Despite the absence of data from animal studies, it is evident that such a reagent is a potential carcinogen. It must thus be handled with great care. This also applies to some other "superpowerful" methylating and ethylating agents such as ethyl fluorosulfonate ("magic ethyl"), methyl trifluoromethanesulfonate (methyl triflate), and Meerwein reagents (trimethyl- and triethyloxonium tetrafluoroborates) (these latter two reagents have the advantage of being solids).

Among the nonsulfated esters having alkylating properties, alkyl phosphates (trimethylphosphate) are mutagenic and carcinogenic in animals.

Inorganic compounds.   Many inorganic compounds are known to be carcinogenic in human beings. These include asbestos, arsenic and its derivatives, chromium(VI), nickel, and their derivatives. Results are less clear with beryllium, cadmium, and especially with lead, for which many authors have refuted its carcinogenic activity. Various other metals (cobalt, iron, antimony, manga-

$$F - \overset{\overset{\displaystyle O}{\|}}{\underset{\underset{\displaystyle O}{\|}}{S}} - O - CH_3 \qquad F - \overset{\overset{\displaystyle O}{\|}}{\underset{\underset{\displaystyle O}{\|}}{S}} - O - CH_2CH_3 \qquad CF_3 - \overset{\overset{\displaystyle O}{\|}}{\underset{\underset{\displaystyle O}{\|}}{S}} - O - CH_3$$

Methyl fluorosulfonate   Ethyl fluorosulfonate   Methyl triflate

$(CH_3)_3OBF_4$ $(CH_3CH_2)_3OBF_4$

Trimethyloxonium   Triethyloxonium
tetrafluoroborate   tetrafluoroborate

$(CH_3O)_3P = O$

Trimethylphosphate

nese, selenium, etc.) form derivatives which are carcinogenic in animals. Metals may be carcinogenic under diverse forms: particles (bronchial cancers due to metallic dust), inorganic derivatives (oxides, sulfides, salts, etc.), metal carbonyls (nickel carbonyl, etc.) and organometallic compounds (tetraethyl lead, cyclopentadienyl derivatives, etc.). It should also be noted that certain inorganic substances (Al, Co, Mn, Ni, Zn, Se, etc.) possess antitumor properties.

Among the common laboratory inorganic chemical substances having carcinogenic properties may be cited derivatives of hexavalent chromium (chromium trioxide, chromates, dichromates, etc.), arsenic (oxides, arsenates III and V, etc.) as well as certain derivatives of nickel, cadmium, and so on. Asbestos is found in various protective materials and it appears that other fibrous substances are now suspected of having definite carcinogenic activity.

Safety and preventive measures in the manipulation of carcinogenic, mutagenic, and teratogenic substances

In order to prevent risks to the health of personnel whose work implicates the use of one or several carcinogenic substances, a certain number of safety and hygiene measures must be applied which take into consideration the various circumstances under which contamination may occur.

The recommendations that will be developed are applicable to all genotoxic substances. The use of potentially genotoxic products must rigorously comply with all hazardous material regulations that must be enforced during the entire course of the operation.

It is the responsibility of the scientific or technical hierarchy (the laboratory director or supervisor) to inform their personnel concerning correct operating procedures and to make available to them all the means of protection necessary for the procedures anticipated. In general, their involvement in improving working conditions is a major consideration in furthering prevention. The latter must always, within reasonable limits, be founded on the mutual consensus of all the personnel involved. For more detailed procedures, see OSHA's Laboratory Standard (29 CFR 1910.1450)* and OSHA's Hazard Communication Standard (29 CFR 1910.1200).

Supplying genotoxic substances.   Supplies of genotoxic substances, obtained either by ordering, from colleagues or by synthesis, must be kept as small as possible and quantities should not exceed those required for the anticipated experiment. Pre-weighed packages of powder prevent the risks of dispersion during weighing, but these are generally more expensive.

Inventory and storage.   Each laboratory (or the central chemical stores, if these exist) must keep an up-to-date register concerning all the potentially genotoxic substances ordered (name of product, origin, grade, quantity, date of reception, dates when used, names of users, etc.) including a Material Safety Data Sheet (MSDS) for each substance.

Genotoxic substances must be stored in specific locations (cupboards, refrigerators, freezers, cold rooms, etc.) and a sign indicating "Hazardous Substances" posted where it can be clearly seen. Flasks and bottles containing genotoxic substances must be placed in hermetically sealed containers or, failing this, in clearly labeled, spill-proof trays. In all cases, the type of material chosen for the containers must be impervious to their contents and to the solvents used to dilute them.

An inventory of all stored carcinogenic substances should be performed regularly and the state of their containers verified.

Information and training.   For each potentially genotoxic substance or family of substances (such as nitrosamines, polyaromatic hydrocarbons, etc.) a document must be drawn up based on the available literature data (IARC monographs, National Toxicology Program documents, etc.). This document, kept up to date as new data appears, must be made easily available for consultation. For substances that are suspected to be genotoxic but for which no toxicological information is available, the same regulations as those applying to carcinogenic substances must be applied.

It is essential that all personnel handling genotoxic substances be informed by their supervisors of the dangers involved and the precautions that must be taken. It is highly desirable that all persons having to handle genotoxic substances follow a specialized course on chemical risks. A worker who is well trained and well informed on safety and who, especially, can recognize and control risks, constitutes a primary element of all effective prevention, particu-

---

*A good resource is *Laboratory health and safety handbook*, by R. Scott Stricoff and Douglas B. Walters. Wiley-Interscience, New York.

larly in the case of the manipulation of genotoxic substances. For many genotoxins, toxic effects are not immediately detectable but appear after a more or less prolonged latency period. This situation can lead to decreased vigilance. Failure to seriously or consistently judge risks can have dire consequences in cases where genotoxic substances are handled.

Work areas.   In laboratories where genotoxic substances are regularly employed, a room should be specially equipped for weighing, dissolving, diluting, manipulating, storing, and eventually destroying them. If possible, this room should be negatively pressurized with respect to the rest of the building. This special room should preferably be located in the least frequented part of the building in order to better control access and avoid superfluous traffic. In this room, as well as on the shelves used to store genotoxic substances, should be clearly posted the warning "Hazardous Substances."

As with all premises reserved for dangerous operations, one person should be made responsible for the hazardous substances room. This person, having sole access to the room's key, should verify that the room remains locked when not in use. Access to this room should be uniquely reserved to personnel directly concerned.

A chart should be posted within the room on which the regulations and guidelines (e.g., Standard Operating Procedures or SOPs, MSDSs) are precisely defined (individual protection, cleaning after use, etc.).

Personal protective equipment.   Before each experiment, the operator should obtain the clothing and protective equipment best adapted to the type of work to be undertaken.

> It is important to be aware of the limitations of all safety equipment and to realize that the choice of this equipment is determined by the type of substance to be manipulated as well as by the nature of the operation.

As in all areas where chemical substances are handled, drinking, eating, smoking, and applying makeup must be prohibited. Chewing the ends of pencils and pens must also be avoided.

Lab coats.

> The wearing of a clean cotton lab coat or disposable coverall is obligatory in all areas where genotoxic substances are handled.

This lab coat, especially reserved for this type of operation, can be of any color and must be worn upon entering the work area and removed upon leaving it. If the lab coat is contaminated in any obvious way, it must be decontaminated or preferably, properly disposed as hazardous waste. Moderately priced, single-use, disposable lab coats are now available, and are preferable to the more permeable cotton lab coat.

Protective goggles.

> As when handling all chemical substances, the wearing of protective goggles is obligatory.

In the case of corrosive genotoxic substances (mustard gases, α-halo ethers, alkyl sulfates, etc.), the use of a face shield is recommended.

Gloves.   Gloves do not assure complete protection and certain genotoxic substances, particularly when they are in solution, can easily penetrate them. Among these genotoxic substances may be cited:

- –halogenated derivatives (1,2-dibromo-3-chloropropane, 1,2-dibromoethylene, etc.)
- –aflatoxins (especially in $CHCl_3$ or DMSO solutions)
- –aromatic amines (especially in methanol solutions)
- –nitrosoamines
- –certain cytostatic agents (Carmustine, etc.)

The type of glove to be used depends on the type of substance and solvent to be handled.

Some vinyl gloves are dissolved by halogenated solvents, in contrast to latex or polyvinyl alcohol gloves. The use of "liquid gloves" or creams is not recommended since, though their effectiveness (of limited duration) has been shown in the case of acids, they have not yet been tested with genotoxic substances. Consequently, avoid the use of creams.

Cotton gloves should be used as the outer pair of gloves to handle electrostatic powders. (Aflatoxins and certain polyaromatic hydrocarbons develop static electricity in the presence of polymeric gloves and may, as a result, disseminate in the atmosphere.) These cotton gloves may be worn over an inner pair of latex gloves.

Some substances such as nitrosodimethylamine (NDMA) and nitrosodiethylamine (NDEA) easily diffuse across rubber surgical gloves.

For handling nitrosoamines, the use of two pairs of dissimilar gloves, is recommended.

After each contamination, the outer gloves should be removed immediately and replaced by a clean pair.

As a general rule, handling doorknobs, faucets, and so on with gloved hands should be avoided. The operations room should not be left before removing gloves, disposing of them in a bag set aside for destruction, and carefully washing hands.

Respirators.   The use of a NIOSH approved dust/mist respirator is recommended when handling powders. However, a respirator equipped with organic vapor cartridges is indispensable when handling highly volatile liquids (halogenated ethers, simple dialkylnitrosoamines, etc.).

Remark: Manipulations on laboratory animals require protection with a cotton face mask, a cap, and cotton overshoes, all of which must be removed upon leaving the experimentation room.

General operations.   As a general rule, overpopulated laboratories should be avoided; each operator should be provided with an adequate working area.

Work surfaces should be easy to clean, uncluttered, and covered with a disposable-type paper or coated with an impermeable surface.

The work area and benches where genotoxic substances are handled should be cleaned by the users after each operation. All manipulations involving pure substances or solutions must be performed within a safety enclosure whose nature depends on the type of discipline (chemistry, biology, etc.), on the type of substance being handled, and on the type of experimentation.

In chemistry laboratories, fume-hoods are mainly used. In general, a fume-hood producing a frontal air current of $100 +/- 20$ linear feet per minute (lfpm) when opened to a reasonable height is sufficient for protection against the majority of toxic substances. These figures take into account air turbulence caused by work-related factors.

Under no circumstances should horizontal laminar flow-hoods be used in biochemistry laboratories. Biology and biochemistry laboratories must instead be equipped with biological safety enclosures that provide for 100% exhaust and 0% recirculation. For genotoxicin substances, only class II enclosures having an air extraction system that expels air to the exterior after it has passed through a total filter and/or an activated charcoal filter are sufficiently effective. Class III enclosures (glove compartment) are completely closed and are reserved for the handling of particularly dangerous genotoxic substances (volatile dialkylnitrosoamines, etc.). The gloves inside a glove compartment are generally always contaminated.

Weighing.   If possible, a balance should be reserved exclusively for genotoxic substances. It should be placed away from all turbulence, preferably inside a vented enclosure. Weighing of substances to an exact, predetermined weight is to be avoided. Instead, a double weighing procedure should be used. Thus, to a previously tared flask is added the approximate quantity of substance desired and the weight is taken. A quantity of solvent necessary to obtain the desired concentration is next added to the flask, which is then carefully stoppered.

Dilutions and sampling.   All operations involving dilution or sampling must be performed under a fume-hood or ventilated enclosure.

Never pipette by mouth.

A propipette or, better, an automated pipette should be used. In contrast to the former, an automated pipette such as "pipette aid," consisting of an electric pump fitted with a filter that retains 0.8 ml of liquid, prevents loss of drops and allows precise distributions of the desired volumes.

Manually regulated, automatic distribution systems equipped with disposable plastic tips allow small volumes of $1-5,000\ \mu l$ to be accurately dispensed.

Labeling.   All flasks containing genotoxic substances (pure or in solution) must be labeled clearly and indelibly with the words "Hazardous Material."

OSHA requires that the label also contain the name of the compound and its hazardous properties.

Decontamination of equipment.   Sampling equipment must be immediately decontaminated either by a specific technique or by use of repeated rinsing. Five successive rinses with quantities of appropriate solvent sufficient to wet the entire surface allows contamination to be reduced to levels close to the limits of detection. If this technique is employed, the used rinse solutions must be collected, a hazardous waste determination made and then disposed as hazardous waste if appropriate.

After each operation, all glassware must be decontaminated by an appropriate technique before being given for washing.

If the floor, workbench, or protective garments are accidentally contaminated, their decontamination by an appropriate method must immediately be undertaken.

Waste storage.   Neither genotoxic substances nor even dilute solutions of these must be thrown down the sink. All wastes must be appropriately disposed.

Solid wastes.   As an operation progresses, the solid wastes that accumulate (toxic substances, Petri dishes, gloves, etc.) are placed in a suitable container for the contents and labeled as hazardous waste as well as with the name of the material and its inherent hazards (e.g., corrosive). Once filled, the container is transferred to the central accumulation area.

Used pipettes and disposable tips are stored in special containers for disposable pipettes while awaiting removal and incineration.

Liquid wastes.   Liquid wastes are poured into a container of suitable material for its contents and labeled as hazardous waste as well as with the name of the material and its inherent hazards (e.g., flammable). This wide-mouthed container should be equipped with a hermetically tight screw-cap. This container should be stored at or near the point of generation under the control of the operator who generated the waste. Once 55 gal of the acutely hazardous waste is reached, then it must be transferred within 3 days to the central accumulation area.

Destruction of carcinogenic substances.   Since carcinogenic agents belong to different chemical families, there is no single, general method for their destruction.

For large quantities of organic substances, in the form of wastes, industrial incineration at temperatures between 1,000° to 1,500°C with a reasonable resident time constitutes the most convenient method of destruction.

In research centers where special furnaces are available, incineration at temperatures higher than 1,200°C is the method of choice. This technique can be applied directly to such articles as Petri dishes containing hazardous wastes such as biohazards. For the destruction of small quantities of hazardous substances, the techniques employed rely on their chemical reactivities.

Alkylating agents, particularly halogenated compounds, may be transformed into nongenotoxic derivatives by nucleophilic agents (water, mineral bases, mercaptans, hyposulfites, etc.). Many direct alkylating agents are efficiently destroyed by overnight contact with a large excess of a 10% aqueous solution of sodium hydroxide, potassium hydroxide, or ammonia. For compounds that are poorly soluble in water, they may first be solubilized in an alcohol (methanol, ethanol). This technique allows the destruction of:

- mustard compounds
- haloethers
- alkyl sulfates
- alkyl sulfonates
- nitrosoureas
- nitrosoguanidines

For indirect alkylating agents, that is, those that require metabolization before becoming active in vivo (e.g., nitrosoamines, hydrazines, etc.), specific methods of destruction will be described.

For the principal families of carcinogenic substances, many selected and verified techniques are regularly published by IARC. Oxidation by potassium permanganate in acid medium may be applied to numerous organic compounds. It should be noted, however, that certain substances such as melphalan or other cytostatic compounds produce highly mutagenic derivatives when this technique is employed. The use of potassium permanganate in alkaline medium allows the destruction of these substances without formation of mutagenic derivatives.

Transporting hazardous substances.   Within the laboratory, solutions of hazardous substances should be transported in leak-proof containers made of material inert to both the substances themselves and their solvents. These containers can in turn be placed in a solid secondary container filled with a quantity of absorbant sufficient to absorb the entire volume of liquid present.

If a substance (solid or in solution) is to be transported to another room, it should first be packaged in a leak-proof container which is placed in a leak-proof secondary container together with a sufficient quantity of absorbant (e.g., vermiculite). This container is then placed in another shockproof container before transportation can be effected. All containers must be carefully labeled.

Replacement products for carcinogenic agents

One of the recommendations put forward for the first time by the 24 June 1974 meeting of the International Labor Organization was that:

All efforts should be made to replace carcinogenic agents and substances to which employees may be exposed during the course of their work by noncarcinogenic agents and substances or by less toxic agents and substances.

Of course, in choosing a replacement substance, it is important to verify all the associated parameters such as its area of application, stability, flammability, explosiveness, acute toxicity, long-term toxicity, and so on. For example, powerful alkylating agents such as methyl fluorosulfonate ("magic methyl") or ethyl fluorosulfonate ("magic ethyl") must, if at all possible, be replaced by less dangerous reagents such as the corresponding Meerwein reagents (trialkyl-oxonium tetrafluoroborates). These alkylating agents have the advantage of being solids.

$$
\underset{\underset{O}{\overset{O}{\|}}}{F - S - O - R}
\qquad\qquad\qquad
R_3OBF_4
$$

Alkyl fluorosulfonates   Trialkyloxonium tetrafluoroborates

$R = CH_3$ $R = CH_3$

$R = CH_2CH_3$ $R = CH_2CH_3$

For some of its uses, diazomethane may be replaced by trimethylsilyl-diazomethane, which appears to be less dangerous.

$$H_2C = N^+ = N^- \qquad\qquad (CH_3)_3Si - H_2C = N^+ = N^-$$

Diazomethane   Trimethylsilyldiazomethane

Within the aromatic amine family, benzidine, a potent bladder carcinogen in human beings, is still used for the detection of peroxidases (investigation of blood traces, etc.). Its replacement by 3,3',5,5'-tetramethylbenzidine, a tentatively noncarcinogenic aromatic amine, allows very sensitive peroxidase detection. For other uses of benzidine and its genotoxic 3,3'-disubstituted derivatives (o-toluidine, o-dianisidine, 3,3'-diaminobenzidine), a number of less toxic replacement compounds can also be proposed, such as 4-chloro-1-naphthol.

Fibrous mineral products, such as asbestos, may be advantageously replaced by innocuous, noncarcinogenic (i.e., based on the limited evidence available) ceramic fibers that are commercialized for this purpose.

Benzidine   3,3'-Diaminobenzidine

3,3',5,5'-Tetramethylbenzidine   4-Chloro-1-naphthol

In the area of solvents, replacement by less toxic liquids is generally not difficult, as indicated in Table 1.19.

Solvents known to be teratogenic in animals — for example, amides such as formamide, N-methylformamide, and N,N-dimethylformamide (DMF) — may be replaced by dipolar, aprotic solvents such as N-methyl-2-pyrrolidone (NMP).

Whenever it is technically feasible, the replacement of a carcinogenic substance by a noncarcinogenic one constitutes an important element in prevention and should be systematically sought.

SUSPECT CARCINOGENIC SOLVENTS (carcinogenic in animals)	PROPOSED REPLACEMENT SOLVENT
Benzene	Cyclohexane Toluene (neurotoxic)
Methylene chloride (dichloromethane) Chloroform (trichloromethane) Carbon tetrachloride (tetrachloromethane) 1,2-Dichloroethane Trichloroethylene Perchloroethylene	1,1,1-Trichloroethane (neurotoxic) (in fires, may form phosgene and hydrochloric acid)
1,4-Dioxane	Tetrahydrofuran
2-Nitropropane	Nitroethane
Hexamethylphosphoric triamide (HMPT)	N,N'-Dimethylpropyleneurea (DMPU)

Table 1.19   Principal replacement solvents for genotoxic solvents.

$$H_2N \longrightarrow CHO$$

Formamide

$$H_3C \longrightarrow NH- CHO$$

N-Methylformamide

H3C
  \
   N — CHO
  /
H3C

N,N-Dimethylformamide (DMF)

N-Methyl-2-pyrrolidone (NMP)

Medical surveillance

There is at this time no simple, efficient method for measuring eventual exposure to a genotoxic substance. In order to better protect workers, biological exposure indices (BEI) have recently been defined. These are warning thresholds corresponding either to a biological reaction to a chemical substance or to concentrations of a chemical substance or its metabolites in tissues, biological fluids, or exhaled air. Industrial hygiene's threshold limit values (TLVs), published by the ACGIH, consist of values based on evaluations of workers' exposure to toxic substances via inhalation and/or skin absorption. Many publications concerning the prevention of profession-related cancers include lists of substances for which appropriate precautions must be taken. Specific publications deal with bladder carcinogens in humans (4-aminobiphenyl, benzidine, 2-naphthylamine, 4-nitrobiphenyl, etc.).

In contrast, *biological surveillance* evaluates exposure to the totality of chemical substances present in the working environment, as quantified by the BEI. In the case of alkylating agents, studies are presently being conducted to determine trustworthy parameters for the biological surveillance of workers exposed to these substances (e.g., determination of mercapturic acids in urine). In particular, the level of hemoglobin and DNA alkylation may, for some alkylating agents (e.g., ethylene oxide, etc.) be correlated with the degree of exposure to these substances in the work environment. Immunochemical techniques using monoclonal antibodies allow detection and quantification of DNA lesions; however, validation of this method still poses difficulties.

While waiting for these types of biological surveillance techniques to become easily accessible,

It is necessary that, in addition to routine analysis of the workplace atmosphere, systematic medical examinations (regular checkups and blood tests) of all personnel (permanent and temporary) in contact with genotoxic substances be performed.

In case of an incident or an accident involving this type of substance, the medical services must be immediately alerted and, if necessary, appropriate decontamination must be rapidly effected.

For women of child-bearing age, effective protection against teratogenic substances is strongly advised. Pregnant women must not work with teratogenic substances.

As an example, many alkylating agents such as nitrosoureas are transplacental carcinogens.

> In summary: Whenever technically feasible, a carcinogenic substance must be replaced by a noncarcinogenic substance. If this is not possible, then all the preventive measures adapted to the particular nature of the risk involved must be applied. These preventive measures will often vary as a function of the substances employed as well as the type of operation effected.
>
> If possible, all operations involving hazardous substances should be performed in the same place. This hazardous substance (e.g., carcinogenic, etc.) operations room must be appropriately indicated and must only be used by personnel having obtained proper training and who are well aware of the risks involved.
>
> It is the research supervisor's responsibility to make sure that their personnel are aware of the risks. They must, moreover, ensure that safety regulations applying to the handling of genotoxic substances are respected.
>
> In all places where hazardous substances are regularly handled, training programs must be established (e.g., as per OSHA's hazard communication standard) and made available to all personnel in contact with these hazardous substances. Periodically updated material safety data sheets regarding the proper handling and disposal as well as the physical and toxicological properties of the chemicals must also be made available.
>
> No easily utilizable biological surveillance test is currently available for persons exposed to hazardous substances in their place of work. Research should be encouraged in this area. As for all those routinely handling toxic substances, regular medical surveillance is indispensable. Moreover, in order to better evaluate the toxicological risks associated with the manipulation of mutagenic, carcinogenic and teratogenic substances, epidemiological surveys of personnel in contact with chemical products should be pursued.
>
> In order to diminish the risk of cancer or malformations in offspring, it is essential that both individual and collective preventive measures be strictly applied.

1.3.3.4 Teratogenic Substances

A teratogenic substance may be defined as one which acts preferentially on an embryo at precise stages of its development, leading to one or more anomalies and, eventually, malformations. Teratogenic substances may also serve to increase the frequencies of certain spontaneous malformations. Teratogenicity must not be confused with embryotoxicity or fetotoxicity, which represent the aptitude of substances to provoke toxic symptoms in the embryo or the fetus and which can eventually lead to their deaths.

Figure 1.14 summarizes the principal steps involved in fertilization to birth, that is, the development stages of the embryo and of the fetus.

Two essential periods can be distinguished:

–*an embryonic period* going from conception and implantation (nesting) to the seventh week and which corresponds to formation of the various organs (organogenesis).
–a *fetal period* going from the seventh week until birth; this period corresponds to the maturing and perfecting of the organs formed during the preceding period.

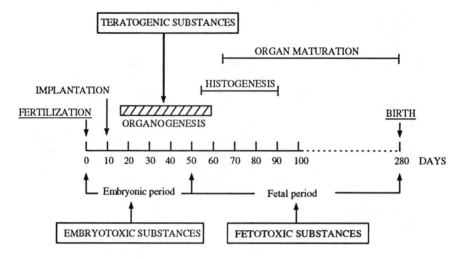

Figure 1.14   Principal stages of embryogenesis.

Mechanisms of embryogenesis perturbation

Perturbations in the development of the embryo and of the fetus may result from exposure, before conception, of the sex cells of one of the parents to genotoxic substances. During gestation, exposure of the mother to embryotoxic or fetotoxic substances can lead to death of the embryo by spontaneous abortion or to problems in the child as indicated by a below-average weight at birth (retarded intrauterine growth) or by the appearance of malformations or functional disorders.

Among the many factors that may influence the teratogenic potency of a substance, the following may be cited:

The development stage of the embryo.   During the first 15 days after conception, the cells are mainly undifferentiated and their exposure to a toxic substance at this time leads either to embryonic death and spontaneous abortion at the start of pregnancy or to a delay in the ulterior development of the fetus, though without the appearance of malformation.

> It is during the 15th to the 60th day, corresponding to the period of organogenesis, that the embryo is most sensitive to the action of teratogenic substances.

Each organ develops according to a precise schedule and any interaction, even for a short period, between a substance and embryonic cells may lead to malformations of the corresponding organ. Table 1.20 shows the differences in sensitivity of the principal organs to the action of teratogenic substances.

During the fetal period (8th to 36th week), the organs, having been formed, are no longer subject to major morphological anomalies, but may suffer from minor malformations and dysfunctions. The nervous system and genital organs, however, continue to differentiate until the end of the fetal period and beyond and are, consequently, much more sensitive to teratogenic substances.

Gestation week

		Embryonic period					Fetal period				Term
1	2	3	4	5	6	7	8	12	16	20-36	38

Division period of the implanted zygote and of the bilaminar embryo

Central nervous system

Heart

Arms

Eyes

Legs

Teeth

Palate

External genital organs

Ears

In general, relatively insensitive to teratogens

Prenatal death	Major anatomic malformations	Physiological anomalies and minor anatomic malformations

Source: Taken from S. Cordier and D. Hemon. *Arch. Malad. Prof.* 1983, **44**(4), 247–51.

Note: Blackened areas indicate the most sensitive periods.

Table 1.20 Stages of embryogenesis sensitive to the action of teratogenic substances.

Thus, the risk of teratogenicity exists at a very early stage of development and continues during the entire gestation with a critical period up to the third month.

The mechanism of action of teratogenic substances. Certain substances act directly as, for example, dimethylmercury or various direct alkylating agents such as nitrosourea. Others require enzymatic metabolization to become toxic (protoxic agents). This appears to be the case for most of the recognized teratogenic substances: thalidomide, polyaromatic hydrocarbons, and so on.

It is probable that at the molecular level, the mechanism of action of numerous teratogens such as alkylating agents involves interaction with DNA, leading to either punctual mutations at the gene level or to aberrations in the structure or the number of chromosomes.

The toxic dose. In contrast to carcinogenic agents for which there are no threshold doses, except perhaps in the case of promoters, there appears to be for teratogenic substances, a threshold below which there is no embryotoxicity and above which a dose-response relationship can be observed. For example, acrylonitrile is teratogenic in rats at an oral dose of 65 mg/kg, but has no effect at 25 mg/kg.

Principal teratogenic substances

While the known teratogenic agents are responsible for the majority of birth defects, numerous, yet unknown environmental factors have an influence on teratogenic potency. This could be the reason why more than 60% of reproductive anomalies remain unexplained. In fact, the distinction between effects due to genetic factors and to teratogenic, xenobiotic agents is not easily established. The genetic constitution of a given subject plays an important role in determining the degree of sensitivity toward toxic agents.

Teratogenic agents are of varied nature. They may be

- physical (ionizing radiation, ultrasonics, etc.)
- microbiological (rubella or herpes virus, etc., toxoplasmosis, etc.)
- chemical (pollutants, drugs, alcohol, tobacco, etc.)

Finally, certain malformations are due to biochemical disorders which occur in the course of embryonic development (e.g., caused by vitamin deficiency or by hypervitaminosis).

Substances known to be teratogenic in animals are classified in the *Registry of Toxic Effects of Chemical Substances* (RTECS) annually revised in the United States by the National Institute for Occupational Safety and Health (NIOSH). Thus, of the 34,000 substances listed in the 1978 edition, 527 are considered teratogenic and, of these, 156 are transplacental carcinogens.

It is, however, surprising that a number of substances classified as teratogens are only weakly toxic; this is the case, for example, of sodium chloride, of galactose, and of lactose. Instead of reproducing this list, it appears more judicious to select compounds whose teratogenic properties have been clearly demonstrated in experimental animals and which are likely to be routinely encountered in chemical and biochemical laboratories. These compounds are listed in Table 1.21.

As this table indicates, a very limited number of substances are actually recognized as teratogenic based on epidemiological studies. These include thalidomide (an antiemetic and a tranquilizer), diethylstilbestrol (a synthetic estrogen), and dimethylmercury (an organic derivative of mercury). Vinyl chloride, a potent human carcinogen, leads, at high doses, to chromosomal anomalies in continually exposed workers. It is thus a suspected teratogen.

A certain number of chemical substances commonly used in the laboratory have been shown to be embryo-toxic and fetotoxic in animals (Table 1.22).

Dimethylmercury   Thalidomide

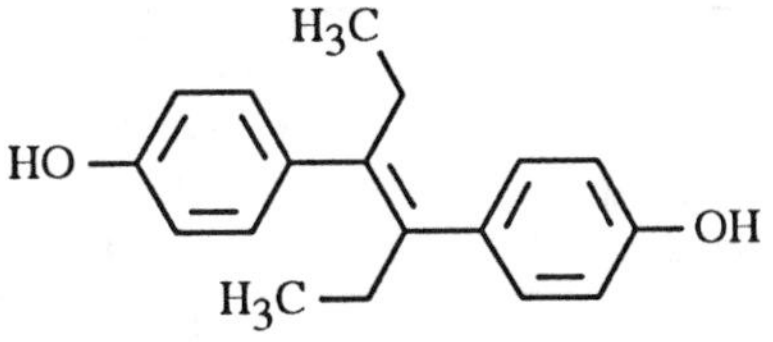

Diethylstilbestrol 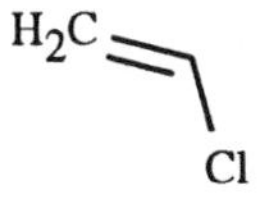   Vinyl chloride

Thus, among the chemical products suspected of affecting embryonic development after paternal exposure may be cited:

–chloroprene
–vinyl chloride
–aromatic hydrocarbons
–lead (inorganic or organic)

Moreover, numerous substances may affect embryogenesis as a result of contact with the mother. Among these are:

–alcohol (ethanol)   –carbon monoxide
–benzene   –lead (inorganic or organic)
–beryllium   –polychlorinated biphenyls (PCBs)
–vinyl chloride   –selenium
–dioxin   –toluene
–formaldehyde   –various other solvents (amides, etc.)
–mercury (inorganic or organic)   –tobacco

Certain professions apparently lead to increased exposure of women to embryotoxic and fetotoxic substances. This is true for industrial chemistry, particularly with respect to laboratory work. In this regard, risks for women working with chemical substances have been estimated as being two to three times higher than in other occupations, though more precise epidemiological studies would be necessary to confirm this estimate.

The relationship between carcinogenicity and teratogenicity

While many chemical substances that are toxic over long periods exert their carcinogenic or teratogenic actions via induction of mutations (punctual or chromosomal), other mechanisms not involving DNA are also possible. Such is the case, for instance, of *transplacental carcinogenesis* in which a carcinogenic substance induces cancer in the descendants of females in contact with this substance. Many substances that are carcinogenic in animal experiments (alkylating agents, vinyl chloride, etc.) are also transplacental carcinogens.

The only documented example in humans is that of DES (diethylstilbestrol) which may lead to the appearance of specific genital cancers (vagina) in the daughters of women treated with this synthetic estrogen. The use of DES in biology is not uncommon, and it should be handled with caution.

| COMPOUND | TERATOGENIC POTENCY | | | |
| | IN ANIMALS (Female) | | IN HUMANS (Female) | |
	Terato-genicity	Targets	Terato-genicity	Targets
Arsenic (As^{3+}, As^{5+})	+	Multiple	?	
Cadmium derivatives	+	Bones	?	?
Nitrogen dioxide (NO$_2$)	+	Multiple	?	?
Dimethylmercury	+	Central nervous system	+	Central nervous system
Selenium	+ (weak)	Eyes Limbs	?	?
Tellurium	+	Brain	?	?
Benzo(a)pyrene	+	Multiple	?	?
Chlorodifluoromethane (Fluorocarbon 22)	+	Eyes	?	?
Chloroform	+ (weak)	Bones	?	?
Vinyl chloride	+ (weak)	Brain	(+)	Central nervous system
Methyl ethyl ketone	+	Bones	?	?
Formamide	+	Palate	?	?
Monomethylformamide	+	Central nervous system, Palate	?	?
Acrylonitrile	+	Bones	?	?
Diethylstilbestrol (DES)	+	Vagina	+	Vagina
Thalidomide	+	Bones	+	Bones

Table 1.21   Principal teratogenic substances.

Compound	Embryotoxicity	Fetotoxicity
Benzene	-	+ (weak)
Chloroprene	+	+
Vinylidene chloride	+	+
Dimethylformamide (DMF)	+ (weak)	
Epichlorohydrin	+	?
Mercury (salts)	+	+
Methyl *n*-butyl ketone	+	-
Carbon monoxide	?	+
Ethylene oxide	+	+
Carbon disulfide	+	+
Tetrachloroethylene	+	+
Carbon tetrachloride	+ (weak)	+
Thallium (salts)	?	+
Toluene	?	+
Xylenes	+	+

Table 1.22   Principal embryotoxic and fetotoxic substances in animals.

Preventive measures in the handling of teratogenic substances

The handling of teratogenic substances is subject to the same rules as those described for mutagenic and carcinogenic substances. Though women of child-bearing age are certainly very sensitive to the action of teratogenic, embryotoxic, and fetotoxic substances, the harmful action of certain chemical substances on the reproductive capacities of humans must not be neglected.

Assuring that pregnant women are not exposed to teratogenic substances is insufficient since this does not guarantee that there is no exposure during the first weeks of pregnancy, the most sensitive period for the embryo.

As a general rule, women of childbearing age and those breast-feeding must be particularly protected from toxic substances (solvents, mercury, etc.).

Potentially teratogenic substances, just as carcinogenic substances, should be the object of very strict regulations.

1.3.4 Classification of Toxic Substances

Toxic substances may be classified according to their mode of penetration into the organism, the type of toxic effect, or their mechanism of action. In the latter case, either the biochemical mechanisms or the state and physicochemical properties of the penetrating substance may be described. This latter presentation has been chosen since it corresponds best to the five categories of substances encountered in the laboratory:

–dusts/fibers
–gases and vapors
–corrosives
–irritants
–solvents

1.3.4.1 Dusts/Fibers

Prolonged inhalation of finely powdered substances such as silica, asbestos, or glass fibers can lead to pneumoconiosis, characterized by an increasing difficulty in breathing. Particles of less than 5 μm are able to accumulate in the alveoli; however, fibers can penetrate alveoli even if their lengths attain 10–20 μm. These particles cannot be rejected since no filtering cilia are present at this level. Dust, taken up by macrophages by way of phagocytosis, is either eliminated or, more often, accumulates in certain regions of the lung.

Amorphous silica is mainly used in laboratories as a chromatographic support (as with silica gel) or for filtrations (Celite, Kieselgunr, etc.). The following values, corresponding to total dust levels, may be cited for various types of silica:

Quartz, fused silica	0.3 mg/m^3 (TLU is 0.1 mg/m^3)
Cristobalite, tridymite	0.15 mg/m^3 (TLU is 0.05 mg/m^3)
Tripoli	see INRS note no. 1468-114-84 for calculations (TLU is 0.1 mg/m^3 of contained respirable quartz)

Asbestos is composed of fibrous magnesium silicate and iron silicate. That utilized in laboratories is generally chrysotile or magnesium phyllosilicate. It has a variety of uses: protective gloves, workbench insulation, heating mantles, and, especially, insulation of distillation columnns, glassware clamps, ovens, Prolonged inhalation of asbestos can lead either to fibrosis, referred to as *asbestosis*, or to cancers (bronchopulmonary, mesotheliomas).

Mineral wool and glass fibers are synthetic silicates. Repeated exposure to these substances may lead to irritations of the respiratory tract, such as laryngitis. Glass fibers are even suspected of favoring the appearance of larynx cancers due to local irritation.

Many substances used in the laboratory, such as isocyanates, pentachlorophenol and powdered metals, may provoke asthmatic or eczematoid allergies, and the like.

It should finally be noted that certain particles of easily oxidized substances such as magnesium, zinc powders, or sulfur, may form explosive mixtures in air.

Always use a NIOSH approved respirator equipped with a dust/mist filter cartridge when handling finely powdered substances (HEPA filter may also be necessary depending on the substance).

–Chromatographic plates should be prepared under a fume-hood (weighing, application, drying).

–The dispersion of particles should be avoided when recovering products from chromatographic plates.

Laboratory equipment containing asbestos should gradually be replaced by substances less dangerous to health.*

1.3.4.2 Gases and vapors

Depending on the type of toxic effects produced on the organism, the following classification may be made:

Simple irritants, which provoke an inflammation of mucous membranes on contact. They may notably cause acute pulmonary edemas. Some examples are:

–ammonia
–hydrochloric, sulfuric, nitric, hydrofluoric acids
–acetic acid
–halogens (Cl_2, Br_2, I_2, etc.)
–ozone and nitrogen oxides (NO, NO_2, N_2O_4)
–phosgene
–sulfur dioxide
–formaldehyde

Secondary irritants which, in addition to the effects already mentioned, above, exert a toxic action on the entire organism as, for example, hydrogen sulfide and hydrogen phosphide (phosphine) which are neurotoxic.

Simple asphyxiating gases which act mainly by decreasing oxygen levels. For a single exposure, they are only dangerous at high concentrations corresponding to 20–30% of exhaled air. Gases fitting this category include nitrogen, hydrogen, carbon dioxide, methane, ethane, and the so-called "inert" gases, argon, helium, and so on.

Chemically asphyxiating gases whose chemical actions prevent oxygen from playing its role. They produce asphyxiation even at very low doses. For example, carbon monoxide binds to hemoglobin while hydrogen cyanide and hydrazoic acid bind to cytochrome oxidase (the terminal enzyme in the mitochondrial respiratory cycle).

In the laboratory, toxic gases and vapors may be liberated during the course of certain reactions (Table 1.23).

Recommendations for the handling of toxic vapors or gases

–The equipment to be used should include gas traps. Toxic gases should never be allowed to escape into the atmosphere, even under a fume-hood. Gas alarms should be utilized where available.

---

*Thermal insulation in the laboratory: asbestos and replacement products. L'Actualité Chimique, French Chemical Society, Paris (1984), pp. 75–78.

Substance	Reagent	Toxic Gases Liberated
Azides	Acids	Hydrazoic acid
Cyanides	Acids	Hydrogen cyanide
Arsenic derivatives	Reducing agents	Hydrogen arsenide (arsine)
Selenium derivatives	Reducing agents	Hydrogen selenide
Tellurium derivatives	Reducing agents	Hydrogen telluride
Hypochlorites or chlorates	Acids	Chlorine
Nitrites or nitrates	Acids	Nitric oxide
Nitrites + Dimethylamine	Acids	Dimethylnitrosamine
White phosphorus	Bases	Phosphine
Metal sulfides	Acids	Hydrogen sulfide
Chlorinated solvents	Heat / fire	Phosgene, hydrochloric acid

Table 1.23 Formation of toxic gases or vapors by chemical reaction between certain substances.

- Always work under a properly functioning fume-hood, preferably in a room under negative pressure reserved for operations involving toxic substances.
- Never work alone.
- Alert the other occupants of the laboratory and those of neighboring laboratories. Clearly indicate outside the laboratory that a dangerous operation is in progress.
- If necessary, wear an appropriate respirator fitted with the proper filter cartridge or an SCBA, as appropriate.
- Ensure that all personnel involved in the operation are properly trained, understand the pertinent properties of the chemicals to be used (i.e., from the MSDS), and are experienced in the contingency plan's emergency procedures.
- For the manipulation of hydrogen cyanide derivatives (cyanide, nitriles, cyanohydrins, etc.), have on hand the corresponding antidotes and alert a doctor who can ensure prompt intervention if necessary. Gas monitors exist for cyanides and should be used.

If an incident occurs, rapidly evacuate the premises and activate the facility's contingency plan's emergency procedures.

Certain substances, called lachrimators, strongly irritate eyes and the upper respiratory tract. These substances must not be dispersed in the atmosphere. This is the case, for example, for bromoacetone, benzyl bromide and chloro-acetophenone.

It should be recalled that there is no relationship between odor and toxicity and that certain toxic substances have anesthetic properties. For example, hydrogen sulfide, which can be detected at very low concentrations (less than 1 ppm), paralyzes the olfactory nerves beyond 150 ppm. Many substances are only noted at an odor threshold well above the chemical's TLV.

1.3.4.3 Corrosives

Corrosive substances provoke brutal destruction of cells. They are not, per se, toxic substances since there is no specific action on the functioning of a molecule, a cell, or a tissue. Corrosive substances are often liquids, noncombustible, and highly reactive chemically. Many of these substances have been previously cited in the sections dealing with water-reactive compounds (p. 26ff) and oxidizing agents (p. 22ff).

Strong acids. Strong, concentrated mineral acids easily attack skin and eyes, causing deep lesions. They provoke severe burns of a yellow or brown color and slow-healing necrosis. In the case of contact with eyes, corneal scabbing occurs and blinding is possible if the projections are abundant. Certain acids, such as hydrochloric acid, do not act immediately, allowing the possibility of copious washing of the exposed tissue, an effective treatment.

Certain acids also have dehydrating and oxidizing properties (H_2SO_4 and HNO_3, respectively).

Concentrated sulfuric acid contains sulfur trioxide (SO_3) which is highly irritating to the respiratory organs.

Similarly, nitric acid (HNO_3) emits vapors rich in nitrogen oxides ($NO + NO_2$) which are very dangerous to skin and respiratory organs (breathing difficulties, pulmonary edema).

Remark: There are many types of nitrogen oxides (N_2O, NO, NO_2, N_2O_3, N_2O_4, N_2O_5), but in toxicology, these concern mainly nitrogen oxide (nitric oxide [NO]), a colorless, relatively nontoxic gas (a weak irritant) and nitrogen dioxide (NO_2), a reddish gas consisting of nitrous vapors that are very toxic to the respiratory tract (dyspnea, cough, acute pulmonary edema, etc.). At low temperatures, nitrogen dioxide is a colorless liquid composed of the dimer, dinitrogen tetraoxide (N_2O_4) which dissociates as the temperature is increased, emitting bright red nitrogen dioxide (NO_2) vapors (Figure 1.15).

Long-term exposure to nitrogen dioxide leads to pulmonary disorders (emphysema, etc.) complicated by immunotoxicity which can favor the development of pulmonary infections. Moreover, nitrogen dioxide, in the presence of secondary amines, can form nitrosamines which possess multiple carcinogenic properties.

$$2\,NO + O_2 \longrightarrow 2\,NO_2 \rightleftharpoons N_2O_4$$

nitric oxide   nitrogen dioxide   dinitrogen tetraoxide
(nitrogen oxide)

Figure 1.15 Autoxidation of nitric oxide.

Dinitrogen oxide or nitrous oxide (N_2O), also known as laughing gas, is mainly used as an anesthetic and is not an irritant. In vivo, dinitrogen oxide oxidizes cobalt ($Co^+ \rightarrow Co^{3+}$) present in vitamin B_{12}, leading to inhibition of methionine synthetase with the possibility of anemia over long periods. It is also genotoxic and teratogenic in animals. Table 1.24 summarizes the toxic properties of the principal nitrogen oxides.

Hydrofluoric acid is a very dangerous substance, being both irritating and severely corrosive to skin and mucous membranes and is a bone-seeker. Burns are very long to heal.

Formic acid (HCOOH) and acetic acid (CH_3COOH) are organic acids that are highly corrosive to skin and mucous tissue. They may lead to ulcerations underneath fingernails. Similarly, halogenated acetic acids (monochloroacetic acid, trifluoroacetic acid, etc.) are very corrosive.

The sulfo-chromic mixture rapidly attacks skin and may provoke allergic reactions. Its use should be avoided.

The relative strength of an acid in water (pH) is not necessarily correlated with its corrosive effects on skin or mucous tissue, as Table 1.25 shows.

Strong bases.   The strong mineral bases or caustic alkalis are sodium hydroxide (NaOH), potassium hydroxide (KOH), ammonium hydroxide (NH_4OH), and calcium hydroxide or slaked lime [$Ca(OH)_2$].

Sodium and potassium hydroxide strongly attack cellular constituents by dissolving keratin, hydrolyzing lipids, and degrading proteins. Their destructive action on eyes is particularly serious. Contrary to a generally held belief, the gravity of base-provoked ocular burns is often greater than those produced by the same quantity of acid.

Formula	Official Name (IUPAC)	Trivial Name	Acute Toxicity (Pulmonary)	Long-Term Toxicity
N_2O	Dinitrogen oxide	Nitrous oxide (Laughing gas)	Non-irritant	Anemia Neurotoxicity (Polyneuritis) Genotoxicity Teratogenicity
NO^*	Nitrogen oxide	Nitric oxide	Weak irritant	Cyanosis Methemo-globinemia
NO_2	Nitrogen dioxide		Strong irritant	Emphysema Immunodepression

*Nitric oxide plays an important regulatory role in the organism. It is formed in various cells (macrophages, etc.) by the oxidation of arginine in the presence of an enzyme, NO-synthetase. Nitric oxide, a free radical species ($\cdot N = 0$), acts as a second messenger and is implicated in cardiovascular effects such as vasodilation. It is also a neurotransmitter and functions as a cytotoxic agent of the immune systems.

Table 1.24 Principal nitrogen oxides and their toxic properties.

ACIDS	CORROSIVE EFFECTS	
Increasing acidity	ON SKIN	ON LUNGS
Perchloric acid	++++	++
Sulfuric acid	++++	++++
Hydrochloric acid	++++	++++
Nitric acid	++++	++++
Hydrofluoric acid	++++	++++
Formic acid	++	+
Acetic acid	++	++

+	*moderate irritant*	+++	*superficial tissue destruction*
++	*pronounced irritant*	++++	*deep tissue destruction*

Table 1.25 Comparative corrosive effects of various acids.

All projections of even a single drop of sodium or potassium hydroxide into eyes must be immediately treated.

Eyes must be abundantly washed with running water for at least 15 minutes. The addition of only a small quantity of water onto projections of sodium or potassium hydroxide does more harm than good. This is because their dilution is very exothermic, thereby accelerating the process of tissue destruction.

Skin burns caused by sodium or potassium hydroxide are very exudative, causing first a brownish coloration followed by slow-healing necrosis.

The ingestion of sodium or potassium hydroxide is also very serious. Intense pain is immediately felt first in the mouth, then in the pharynx and digestive tube, leading to diarrhea, bloody vomiting, and a state of shock. Glottal edema may be produced, with the possibility of asphyxiation. The lethal dose is 7–8 g for an adult. Secondary complications are frequent, notably perforations or peritonitis.

The effects of ammonium hydroxide and of slaked lime are less serious. Ammonium hydroxide however, releases highly irritating vapors that can lead to severe respiratory problems such as acute pulmonary edema.

With the exception of quaternary ammonium hydroxides, organic bases such as amines are less caustic than mineral bases. Contact with amines such as dimethylamine, cyclohexylamine, pyrrolidine, and certain nitrogen-containing heterocyclic bases such as pyridine and quinoline, leads to irritation of the skin (dermatitis), eyes (conjunctivitis), and of the respiratory tract (cough, pulmonary edema). Eczematous allergic reactions are also possible (Table 1.12, p. 57).

As Table 1.26 indicates, corrosive effects increase with basicity.

BASES	CORROSIVE EFFECTS
Increasing basicity	
Sodium hydroxide	++++
Potassium hydroxide	++++
Quaternary ammonium hydroxides	++++
Ammonium hydroxide	+++
Calcium hydroxide	+++
Dimethylamine	++
Sodium carbonate	++
Sodium bicarbonate	+

+	*moderate, superficial irritant*	+++	*simple tissue destruction*
++	*pronounced irritant*	++++	*serious tissue destruction*

Table 1.26   Comparative corrosive effects of various bases.

Dehydrating agents.   Because of their affinity for water, many dehydrating agents produce burns when in contact with skin or mucous tissue. Among these dehydrating agents, some, such as sulfuric acid, perchloric acid, and strong mineral bases (sodium and potassium hydroxide), have been previously discussed with regard to their corrosiveness.

Other corrosive dehydrating agents are calcium oxide or lime, thionyl chloride, phosphorus pentaoxide, and certain solvents such as toluene, the xylenes, absolute ethanol, and so on.

Washing under running water is always the best treatment in the case of contact with a dehydrating agent.

Oxidizing agents.   Powerful oxidizing agents are always very corrosive. The most aggressive are concentrated nitric acid, concentrated perchloric acid (less than 72%), chromium trioxide, and concentrated hydrogen peroxide.

Other corrosive substances.   Numerous nonoxidizing, nondehydrating substances are nevertheless very corrosive to skin and mucous tissue. Among these are bromine, fluorine, boron trifluoride (BF_3), chlorosulfonic acid ($ClSO_3H$), dimethyl sulfate [$(CH_3)_2SO_4$], diethyl sulfate [$(C_2H_5)_2SO_4$], titanium tetrachloride ($TiCl_4$), and stannic chloride ($SnCl_4$).

Certain compounds such as phenol (C_6H_5OH) may provoke, in addition to local actions on skin, serious generalized disorders (notably, nervous and renal). Other substances, in addition to being corrosive, are irritants under prolonged exposure. These include phosphorus halides and oxyhalides, methyl bromide, acrolein, formaldehyde (or formol), cobalt derivatives, phenothiazines, piperazine, dithiocarbamates, hydroquinones, mercaptopropionic acid and its derivatives (acrylates and polythiols), and derivatives of methacrylic acid.

Recommendations for the handling of corrosive substances

-Wear goggles, gloves, and, if necessary, a face mask at all times.
-Always work under fume-hoods in good operating condition and verify that they are not deteriorated by the substances handled.
-Avoid working with large quantities.
-All pipetting of corrosive substances (and noncorrosive substances in the vicinity) by mouth must be forbidden.

1.3.4.4 Irritants

Irritants in contact with skin or mucosa (ocular, nasal, respiratory, digestive) provoke more or less intense inflammatory reactions which may be classed as direct intoxications. These inflammatory reactions must not be considered benign since some may lead to death as, for example, in the case of acute pulmonary edemas (APE). Corrosive substances are irritants at low concentrations.

As as been discussed previously with reference to gases and vapors, irritants may be classified into two categories:

Primary irritants. These exert uniquely a local action. For most subjects, the lesions formed are of a similar aspect, their gravity depending on the nature of the substance, its concentration, and the duration of contact.

Secondary irritants. In addition to a local action, these substances affect the entire organism. This is the case, for example, of hydrogen sulfide (H_2S) which, in addition to its effects on mucous tissue (conjunctivitis, respiratory deficiency), is both a neurotoxin, provoking central respiratory depression, as well as a potent enzyme inhibitor, blocking the action of cytochrome oxidase in the respiratory chain.

The effects of irritants vary depending on their site of action.

On the respiratory tract

Certain irritants only act on the upper respiratory tract (nose, throat, pharynx, larynx, trachea) producing rhinitis, laryngitis, or pharyngitis. Sensory irritants act on the cholinergic nerve terminals of the trifacial nerve (mainly by interacting with protein nucleophilic groups but also by cleavage of disulfide bonds). This is the case for the compounds listed in Table 1.27.

Other irritants attack the lower respiratory tract, that is, the bronchi, causing bronchitis and bronchopneumonia, and the lungs, leading to pulmonary edema. These pulmonary irritants exert their effects via activation of the J receptor of the vagal nerve located on the surface of lung alveoli. Among these substances are those listed in Table 1.28.

For certain irritants such as phosgene ($COCl_2$), nitrogen dioxide (NO_2), or ozone (O_3), pulmonary edema manifests itself 4–24 hours after exposure.

Ammonia	NH_3
Ammonium hydroxide	NH_4OH
Acid halides	HF, HCl, HBr, HI
Sulfur oxides	SO_2, SO_3
Formaldehyde	CH_2O

Table 1.27   Upper respiratory tract irritants.

Nitrogen dioxide	NO_2
Phosgene	$COCl_2$
Bromine monochloride	BrCl
Metal carbonyls	$Ni(CO)_4$
	$Fe(CO)_5$
	$Ru(CO)_6$
	$H_2Fe(CO)_4$
Nitro halogenated hydrocarbons	Trichloronitromethane (chloropicrin)
	1,1-dichloro-1-nitroethane
	1-chloro-1-nitropropane
Ozone	O_3

Table 1.28   Pulmonary irritants.

Particular care must be taken with substances that do not produce effects immediately upon exposure, but whose irritating action is delayed. This is the case for yperite (mustard gas) whose toxic effects (blistering in the areas exposed due to its vesicant properties, conjunctivitis, bronchitis) are not observed until several hours after exposure (1–3 hours in general).

On eyes

The irritant may lead to serious watering, as in the case of the following lachrimators:

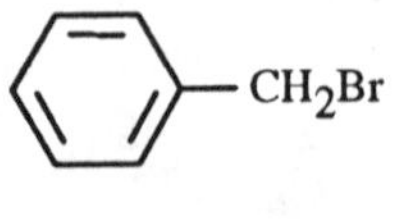

Benzyl bromide

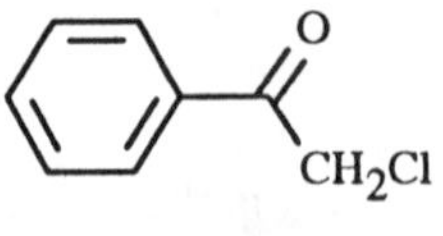

Chloroacetophenone

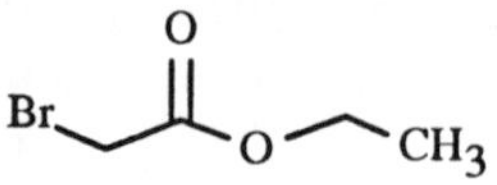

Ethyl bromoacetate

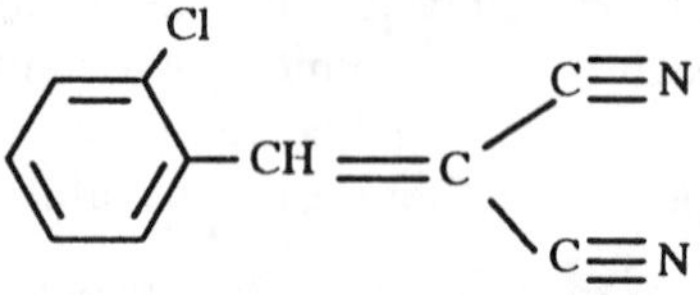

o-Chlorobenzylidene malononitrile

Certain synthetic reactions may liberate such substances and efficient trapping of these must be assured. This is true for the halogenation of carbonyl compounds. Thus, bromination or chlorination of acetone may liberate bromoacetone or chloroacetone, both particularly potent lachrimators.

Vapors emitted by certain strong irritants provoke, in the case of prolonged contact, serious eye damage that may lead to blindness. Thus, osmium tetroxide (OsO_4) or osmic acid, causes eye-watering due to conjunctivitis, followed by a burning sensation with appearance of halos and, ultimately, the possibility of blindness. The mere contact of traces of osmium tetroxide with eyes may lead to intense irritation, often persisting several days after exposure.

On skin

The irritant effect may go from a simple erythema or chemical burn to veritable necrosis (e.g., in the case of a corrosive substance).

Irritation-provoked contact dermatitis is a term which groups different types of modifications occurring to skin, such as burns, erythema, purpura, blisters, eczema, hyperkeratosis, pustules, and even noninflammatory reactions such as dryness and roughness. Two types of cutaneous irritation are usually distinguished:

- *Acute irritation*, which consists of a reversible, local inflammatory response of normal skin to an aggression caused by a single application of a chemical substance without the intervention of immunological mechanisms;
- *Cumulative irritation*, which is reversible and results from repeated or continuous exposure to a substance incapable by itself of producing acute irritation.

The factors influencing an individual's response to a chemical irritant depend both on the substance itself as well as on the individual. The intensity of the irritation depends on:

- the nature of the chemical substance
- the concentration of irritant
- the length of time with which the skin is in contact with the irritant
- the type of skin which comes into contact with the irritant
- individual resistance to aggression by chemical substances; differences in resistance are due to skin pigmentation, age, sex, hygiene, medical treatment, and the presence of skin lesions

Table 1.29 summarizes the increasing irritant potency of various classes of chemical substances.

The majority of solvents can easily dissolve the skin's protective lipid coating and, as a result, many of these are skin and mucous tissue irritants. The degree of irritation due to a solvent increases in going from oxygenated derivatives to saturated hydrocarbons (Table 1.30).

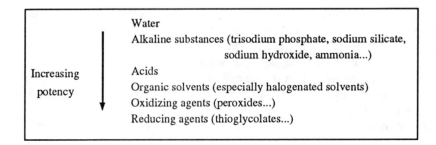

Table 1.29   Increasing irritant potency of different classes of substances.

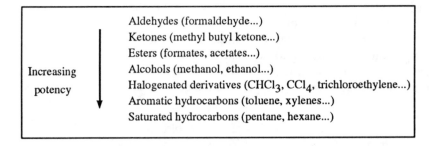

Table 1.30   Increasing irritant potency of solvents.

The biochemical mechanisms involved in irritation phenomena have not been very well characterized. Two mechanisms may be invoked. The first consists of corrosion accompanied by membrane alterations and cell death. The second mechanism corresponds to release of chemical messengers by aggressed cells, which do not necessarily die. The best-known messengers are mediators of inflammation.

Certain irritants also have allergenic properties. Numerous chemical products (organic, inorganic, and organometallic) used in laboratories have irritant properties (Table 1.31).

Remark: As noted in the discussion concerning allergenic substances, some of these substances are also primary irritants. This is the case for various compounds not listed in the table of allergenic substances (Table 1.12, p. 57) such as:

   −alkaline persulfates (ammonium persulfate, etc.)
   −alkaline hypochlorites
   −alkaline sulfites
   −cationic detergents (quaternary ammonium salts, etc.)
   −dicyclohexylcarbodiimide
   −cyanoacrylic resins (cyanoacrylic glues responsible for asthma).

ACIDS, ANHYDRIDES, ACID HALIDES

<u>Inorganic</u>   <u>Organic</u>

- hydrochloric acid   - formic acid
- hydrobromic acid   - acetic acid
- hydrofluoric acid   - chloroacetic acid
- perchloric acid   - bromoacetic acid
- sulfuric acid   - acrylic acid
- nitric acid   - crotonic acid
- phosphoric acid   - lactic acid
                                                       - salicylic acid
                                                       - oxalic acid

- sulfuric anhydride   - acetic anhydride
- sulfurous anhydride   - haloacetic anhydride
- phosphoric anhydride   - propionic anhydride
- chromium trioxide   - butyric anhydride
- arsenic trioxide   - phthalic anhydride
- selenium trioxide
- vanadium trioxide
- osmium tetroxide

- phosphorus halides (PCl_3, PCl_5)   - acetyl halides (acetyl chloride...)
- boron halides (BF_3, BCl_3, BBr_3)
- cyanogen halides (BrCN, ClCN)

- hydrogen sulfide (H_2S)
- phosphine (PH_3)

METAL SALTS

- alkaline chromates
- alkaline dichromates
- mercuric chloride
- zinc chloride
- hexachloroplatinates

Table 1.31   Principal inorganic and organic irritants.

ORGANIC HALOGENATED DERIVATIVES

Haloalkanes   Haloalkenes

- methyl halides   - vinylidene chloride
- methylene chloride   - allyl chloride
- 1,2-dichloroethane   - chloroprene
- 1,1,2,2-tetrachloroethane   - trichloroethylene
- benzyl halides   - perchloroethylene

Haloarenes

- chloronaphthalenes
- polychlorinated biphenyls (PCBs)
- polybrominated biphenyls (PBBs)
- dioxin (TCDD)

BASES, BASIC SALTS

Inorganic   Organic

- sodium hydroxide   - ethylamine
- potassium hydroxide   - dimethylamine
- ammonium hydroxide   - trimethylamine
- calcium hydroxide   - butylamines
- alkaline carbonates   - allylamine
- alkaline phosphates   - cyclohexylamine
- alkaline silicates   - ethylenediamine
- alkaline hypochlorites
                                         - amino alcohols (ethanolamine)
                                         - bases (pyridine, quinoline)
                                         - cationic detergents
                                         - hydroxylamines
                                         - hydrazines

Table 1.31 (continued)   Principal inorganic and organic irritants.

ORGANIC OXYGEN DERIVATIVES

Alcohols, Phenols, Hydroperoxides, Peroxides		Esters
- alcohols :	butanols	- methyl formate
	allyl alcohol	- ethyl formate
	benzyl alcohol	- ethyl chloroformate
	cyclohexanol	- methyl acetate
	amyl alcohols	- ethyl acetate
- haloalcohols		- ethyl chloroacetate
- phenols :	phenol	- methyl acrylate
	cresols	- dimethyl sulfate
	diphenols	- diethyl sulfate
	triphenols	- trimethyl phosphite
- halophenols		- tributyl phosphate
-*t*-butyl hydroperoxide		
- benzoyl peroxide		

Aldehydes, Acetals	Ethers
- formaldehyde	- diisopropyl ether
- acetaldehyde	- dibutyl ether
- chloroacetaldehyde	- dioxane
- acrolein	- diethyl ether
- crotonaldehyde	- epichlorohydrin
- glutaraldehyde	- vanillin
- furfural	- haloethers
- methylal	- epoxy resins

Ketones, Quinones

- methyl butyl ketone	- isophorone
- methyl amyl ketone	- α-haloketones
- acetonylacetone	- p-benzoquinone
- mesityl oxide	- haloquinones

Table 1.31 (continued)   Principal inorganic and organic irritants.

HYDROCARBONS

Saturated hydrocarbons   Aromatic hydrocarbons

- n-alkanes   - toluene
- fuels   - xylenes
- white spirit   - mesitylene
- kerosene   - styrene
                                          - biphenyl

Ethylenic hydrocarbons

- cyclohexene
- turpentine

ORGANIC SULFUR DERIVATIVES

- 2-mercaptoethanol
- yperite (mustard gas)
- carbon disulfide

ORGANIC NITROGEN DERIVATIVES

Nitrated derivatives   Other nitrogen derivatives

- nitroethane   - diethylformamide
- nitropropanes   - acrylamide
- nitrobutanes   - acrylonitrile
- chloropicrin

Table 1.31 (continued)   Principal inorganic and organic irritants.

1.3.4.5 *Solvents*

Importance and uses of solvents

Among the continuously increasing number of chemical compounds synthesized (more than 8 million) and exploited (approximately 70,000) by human
beings, solvents, despite their minor numerical importance (several hundred),
play an essential role in industry, in laboratories, and in everyday life.

Solvents have multiple uses. They may be employed:

−for dissolving
−for extracting
−as reaction media
−for purification (crystallization, chromatography)

−for cooling
−as reagents or starting material (benzene, 1,2-dichloroethylene, carbon tetrachloride, etc.)
−for washing glassware
−as cleaning agents
−for special effects

General precautions in the use of solvents

−Always wear safety goggles. Chlorinated derivatives (chloroform, methylene chloride, etc.) are dangerous to eyes. The wearing of contact lenses may aggravate irritation.
−Choose solvents carefully; avoid the most toxic ones, especially those having carcinogenic (benzene) or teratogenic (formamide, etc.) properties.
−Never pipette even small quantities by mouth. Certain substances are corrosive (pyridine, carboxylic acids, etc.), while others have cumulative effects (methanol, carbon disulfide, halogenated derivatives, etc.).
−Avoid skin contact. Most solvents dissolve lipids, which favors the appearance of dermatitis. Others may produce allergies (pyridine, etc.) while some solvents (dimethylsulfoxide, etc.) may facilitate the cutaneous penetration of toxic substances.
−If possible, always work under an efficient fume-hood in order to avoid air dispersion.
−Stopper all solvent-containing flasks and bottles after use.
−Always work far from heat sources, hot spots, or sources of static electricity (especially in the case of flammable solvents).
−Do not store overly large quantities of dangerous (chloroform, non-stabilized ethers, carbon disulfide, etc.) or flammable (ethers, ketones, esters, etc.) solvents in the laboratory. Use less than 5 l flasks if these are shock-resistant and less than 1 l if they are not.
−Avoid halogenated solvents whenever possible. Fire and/or heat can form phosgene and hydrochloric acid.
−Do not dispose of any solvents down the sink.
−Recover used solvents, taking care to separate halogenated from non-halogenated ones (different costs for destruction).

Principal risks

Risks related to flammability.   These risks are discussed in this chapter on risks related to physicochemical properties.

Risks due to instability.   Among the common solvents, only carbon disulfide must be considered a particularly unstable substance. Due to its low ignition point (100°C), carbon disulfide may spontaneously ignite in air by simple contact with a hot surface. Moreover, this solvent is particularly explosive when mixed with air and has a very broad explosive range (LEL = 1.25%; UEL = 50%).

Certain solvents, especially ethers, can react with oxygen in air to form unstable peroxides. The latter may cause solvents to explode violently during the course of distillation or evaporation. Always date ether cans when first opened and dispose of on a routine basis if not promptly used.

It is imperative that the presence of peroxides be tested for in all solvents prone to autoxidation.

The peroxide test: To 10 ml of solvent, add 1 ml of a freshly prepared 10% aqueous solution of potassium iodide. The presence of peroxides is indicated by the appearance of a persistent yellow coloration (due to the formation of iodine). The addition of several drops of acid facilitates the detection.

In the case of a positive test, the peroxides must be eliminated either by treatment with an acidic, aqueous solution of ferrous sulfate, by the action of lithium aluminum hydride, or by passage through a column of activated alumina.

Easily peroxidized solvents:

–Diethyl ether (ethyl ether)
–Diisopropyl ether
–Tetrahydrofuran
–Tetralin
–Decalin
–Cumene
–Dioxane
–Monoglyme
–Methyl isobutyl ketone
–Isobutyl alcohol (2-butanol, etc.)

After purification, peroxide-free solvents may be stored over sodium pieces or, better, over molecular sieves, away from light and under an inert atmosphere.

Risks related to toxicity. In addition to the previously cited risks, many organic solvents present health hazards (neurotoxicity, hepatotoxicity, nephrotoxicity, carcinogenicity, skin problems, fetal toxicity, teratogenicity, effects on the reproductive system, etc.).

The evaluation of the maximum permissible exposure to risk requires consideration of both the physiological activities and the physical properties of the solvents concerned (boiling point, flash point, spontaneous ignition point, vapor pressure, as well as explosive properties). As an example, a toxic solvent having low vapor pressure is less dangerous under the same conditions than another solvent of similar toxicity but having higher vapor pressure.

While it is difficult to determine a pathology common to all solvents, they nevertheless share a certain number of types of toxicity, such as:

–central, and sometimes peripheral, neurotoxic actions

- −mucocutaneous effects
- −digestive, hepatic, and/or renal disorders
- −effects on the reproductive system

In the work environment, simultaneous exposure to several solvents is common and their additive toxicities (synergy) present nonnegligible aggravating factors. For example, the neurotoxicity of hexane is greatly increased in the presence of methyl ethyl ketone.

Acute intoxication. The high affinity of solvents for lipid-rich (phospholipids, etc.) organs is mainly responsible for the appearance of acute intoxications affecting the liver, kidneys, and other organs. In this toxic process, the molecules act without requiring prior biotransformation.

- Neurotoxicity. For most solvents, the attack of the nervous system begins by an excitatory phase (euphoria, headache, etc.) followed by a depressive phase (anesthesia, sleepiness, loss of consciousness, more or less deep coma) which may lead to death. Upon coming out of a coma, a prenarcotic syndrome may appear (drunkenness, sleepiness, dizziness, fatigue) which may dissipate only slowly and which, by diminishing vigilance, may compromise the intoxicated person's safety. In most cases, however, removal of the victim from the intoxicating environment followed by reanimation allows rapid recovery with no lasting side effects.

- Cardiotoxicity. Some solvents (chloroform, trichloroethylene, etc.) may trigger myocardial hyperexcitability, interfering with efficient pumping of blood. Cardiac deficiency (ventricular fibrillation) may occur, explaining, for example, the brutal death of drug-abusers having breathed massive doses of solvents. Persons having cardiovascular problems as well as heavy smokers are much more sensitive to the cardiotoxic effects of solvents and they must be particularly vigilant.

- Effects on skin and mucosa. Many solvents are caustic to skin and mucosa (acetic acid, aliphatic amines, pyridine, etc.) and may lead to more or less serious burns. Similarly, many polychlorinated* (methylene chloride, chloroform, trichloroethylene, etc.) and nitrated (2-nitropropane) derivatives produce intense irritation when they come into contact with skin or eyes.

Long-term intoxication. Prolonged exposure to sometimes even very small doses of liposoluble solvents leads, over a more or less long period, to often irreversible effects on numerous organs (nervous system, liver, kidneys, bone marrow, etc.). The disorders produced are very difficult to detect and require attentive medical surveillance.

Depending on an organ's capacity to transform xenobiotic substances such

---

*Carelessly stored (air, light, etc.), nonstabilized halogenated solvents decompose to release highly irritating acutely toxic substances (HCl, phosgene, etc.) whose presence aggravates burns produced by contact of these solvents with skin and, especially, eyes (keratitis, conjunctivitis, etc.).

as solvents, the manifestations of long-term intoxication will present a variable selectivity. Some disorders are reversible (liver and kidney disorders), while others are irreversible (aplastic anemia, cancers, etc.).

- Effects on the central nervous system. The majority of solvents are toxic to the central nervous system over long periods, though with highly variable intensities. Among these solvents, some, such as toluene, trichloroethylene, and carbon disulfide, may, in the case of severe intoxication, trigger the appearance of a psychoorganic syndrome characterized by:

  −amnesia (decreased attention and memory, etc.)
  −diminished intelligence (decreased reasoning powers)
  −affective and personality disorders (irritability, fatigue, decreased libido, depressive tendencies, etc.)

  To this list may often be added neurovegetative disorders characterized by abnormal perspiration, palpitations, dizziness, digestive disorders, and so on. These same solvents may also attack the central nervous system more seriously (brain atrophy). Thus, toluene may produce degeneration of the brain and cerebellum.

- Effects on the peripheral nervous system. A few solvents (hexane, methyl butyl ketone, trichloroethylene) can seriously affect the peripheral nervous system (polyneuritis, etc.). Methanol has a particular affinity for the optic nerve and often produces total blindness.

- Effects on liver and kidneys. As with most liposoluble xenobiotic substances, solvents are metabolized by the liver and eliminated by the kidneys. These organs are thus major targets of these compounds. For example, polyhalogenated solvents (carbon tetrachloride, chloroform, 1,1,2,2-tetrachloroethane, etc.), 2-nitropropane, and N,N-dimethylformamide produce serious necrosis of liver (jaundice) and kidneys (nephritis). Subjects having hepatic disorders (jaundice, cirrhosis, hepatitis, etc.) are much more sensitive to the hepatotoxic effects of these solvents.

- Effects on skin and mucosa. Repeated contact of various solvents (halogenated solvents, amines, amides, pyridine, etc.) with skin or eyes often leads to dermatitis. A few substances (turpentine, etc.) produce allergic contact dermatitis (contact eczema).

- Effects on blood. Many nitrogenated solvents (nitroalkanes, nitrobenzene, aniline, N,N-dimethylaniline, etc.) transform hemoglobin (red blood cell pigment) into methemoglobin, which is unable to assure oxygen transport. This effect leads to cyanosis (bluish coloring of skin and mucosa) followed by more or less pronounced hemolytic anemia. These solvents are particularly dangerous since they penetrate skin very easily.

  Moreover, chromosome mutations (clastogenic effects) have been observed in the lymphocytes of workers in contact with various solvents (benzene, toluene, trichloroethylene, etc.) but no clear-cut correlation has yet been established between the degree of exposure and the persistence of these chromosomal aberrations.

• Carcinogenic potency.   Of all solvents, only benzene is a recognized carcinogen in humans and its use should be seriously controlled in order to avoid cases of benzene poisoning, still very numerous. Benzene can, in fact, be considered "public enemy number 1." Repeated exposure to even minimal doses may lead to sometimes irreversible bone marrow damage, resulting in aplastic anemia (absence of red blood cell formation) and sometimes in leukemia.

   Moreover, many other solvents have been shown to be carcinogenic in animals and must, as a result, be considered potentially carcinogenic in humans and handled as such (Table 1.32). A few epidemiological studies seem to indicate a greater incidence of pancreatic and blood cancers (lymphomas, Hodgkin's disease) among subjects exposed to these solvents.

• Embryotoxic properties.   While only a few solvents (chloroform, etc.) have been shown experimentally to be embryotoxic in animals, many (toluene, xylenes, benzene, etc.) have been observed to slow fetal growth. Nevertheless, certain solvents (methyl ethyl ketone, formamide, methyl and ethyl ethers of ethylene glycol or cellosolves, N-methylformamide) are teratogenic in animals.

   Because most solvents cross the placental barrier with ease, *pregnant women must exercise extra caution.* Recent epidemiological studies have in fact shown a higher rate of birth defects (especially with respect to the nervous system) in children of women having had prolonged exposures to solvents, however, generalizations are, for the time being, premature in this

SOLVENT	ANIMAL	TUMOR LOCATION
Methylene chloride	rat, mouse	liver
Chloroform	rat, mouse	kidneys
Carbon tetrachloride	rat, mouse	liver
1,2-Dichloroethane	rat, mouse	variable
1,1,2,2-Tetrachloroethane	rat, mouse	liver
Trichloroethylene*	mouse	liver
Tetrachloroethylene	mouse	liver
1,4-Dioxane	rat	nasal cavity, liver
2-Nitropropane	rat	liver
N,N-Dimethylformamide	rat	gastro-intestinal tract
Hexamethylphosphoric triamide	rat	nasal cavity

*For some authors, the carcinogenic action of trichloroethylene is due to stabilizing agents present in technical grade solvent. Even with pure trichloroethylene, however, hepatic cancers have been observed in certain strains of mice.

Table 1.32   Principal carcinogenic solvents in animals.

regard. There is reason to believe that prolonged contact with certain solvents promotes spontaneous abortions in women.

Principal dangerous solvents

Hydrocarbon solvents.   Among the saturated hydrocarbons (alkanes) only hexane, after prolonged exposure, affects peripheral nerves leading to polyneuritis (lower limb paralysis), the gravity of which depends on the duration of exposure. This does not seem to be the case for pentane and heptane, which are good replacement solvents for hexane.

In male rats, isooctane and decalin have nephrotoxic properties.

Carcinogenic in humans, benzene is by far the solvent presenting the greatest health risk.

Benzene has a selective, cumulative, and often irreversible effect on bone marrow, where most blood cells (red and white cells, platelets) are formed, leading to the appearance of aplastic anemia (absence of red blood cell regeneration) and/or leukemia.

Pure toluene and xylenes, while exhibiting none of this type of myelotoxicity, are more potent neurotoxins than benzene. Long-term exposure to these aromatic hydrocarbons can provoke irreversible lesions of the central nervous system (encephalitis, etc.), affecting psychophysiological functions (insomnia, behavioral disorders, etc.).

Benzene, toluene, and xylenes are all teratogenic in embryos, producing skeletal anomalies and significantly retarded growth in animals. Cyclohexane has none of the inconveniences of these aromatic solvents and may, in most cases, be an acceptable substitute.

Halogenated solvents.   In general, halogenated solvents are severely toxic to the nervous system and, sometimes, the heart. Among these, carbon tetrachloride, 1,1,2,2-tetrachloroethane, chloroform, and 1,2-dichloroethane are potent hepatotoxins and are often dangerous to kidneys. Like methylene chloride,* these solvents are hepatic carcinogens in rats and mice. They may be replaced by 1,1,1-trichloroethane (Table 1.33), but a nonchlorinated solvent is preferable as a replacement.

Chlorinated ethylene derivatives (trichloroethylene, tetrachloroethylene, or perchloroethylene) are highly toxic to the heart and nervous system. The study of their carcinogenicity in animals and humans (epidemiological surveys) has not yet been completed. However, they may be considered as weak carcinogens whose replacement in industry (metallurgy, etc.) poses problems.

Table 1.33 summarizes the toxicological data concerning a few chlorinated solvents.

---

*Methylene chloride is considered to be relatively nontoxic to liver and kidneys, but has a very corrosive effect on skin (eczema) and eyes (conjunctivitis, etc.). Recent studies by the National Toxicology Program in the United States indicate that at high doses, methylene chloride produces liver cancer in rats and mice. These results have recently been confirmed by the IARC in Lyon.

SOLVENT NAME AND FORMULA	ORGANOTOXICITY					CARCINO-GENICITY IN ANIMALS	GENERAL TOXICITY
	Nervous system	Liver	Kidney	Heart	Skin Eyes		
Methylene chloride* CH_2Cl_2	++	-	-	+	+	*	Moderate
Chloroform* $CHCl_3$	+++	++	+	+++	+	*	High
Carbon tetrachloride* CCl_4	+	+++	+++	++	+	*	High
1,2-Dichloroethane* $ClCH_2CH_2Cl$	++	++	+	-	+	*	High
1,1,1,-Trichloroethane (Methylchloroform) Cl_3CCH_3	+++	+	-	-	+	O	Low
1,1,2,2-Tetrachloroethane* $Cl_2CHCHCl_2$	+++	+++	++	+	+	*	High
Pentachloroethane (pentaline) Cl_3CCHCl_2	+++	++	+	-	-	*	High
1,2-Dichloropropane $CH_3CHClCH_2Cl$	++	++	+	-	-	O	Moderate
1,2-Dichloroethylene $ClCH=CHCl$	+++	$\pm$	$\pm$	+++	+	*	High
Trichloroethylene* $Cl_2C=CHCl$	+++	+	+	+++	+	*	Moderate
Tetrachloroethylene* $Cl_2C=CCl_2$	+++	+	-	+	+	*	Moderate
Chlorobenzene	+++	+	+	-	+	O	Moderate

* Recognized as a cause of occupational disease

Organotoxicity : + low ; ++ moderate ; +++ high

Table 1.33 Comparison of the toxicities of the main chlorinated solvents.

Other solvents.   Of all the alcohols, methanol is by far the most dangerous. It is a cumulative toxin that exerts selective action on the optic nerve, diminishing visual acuity and often leading to blindness.

Among the ethers, 1,4-dioxane, besides being a skin, eye, and respiratory tract irritant, also possesses hepato- and nephrotoxicity. It produces cancers (nasal cavity, liver) in rats.

Glycol ethers and acetates (cellosolves), which are widely employed solvents, exhibit variable toxicity depending on the length of the alkyl chain. The monomethyl ether of ethylene glycol (methylcellosolve or methoxyethanol, $CH_3OCH_2CH_2OH$) as well as the monoethyl ether (ethylcellosolve or ethoxyethanol, $CH_3CH_2OCH_2CH_2OH$) are toxic to bone marrow and lymph organs (hemolytic anemia, white blood cell cytopenia), to the central nervous system (encephalopathy, intellectual deterioration, etc.), and to the reproductive system in humans (atrophy of testicles, azoospermia). Methoxyacetic acid (CH_3OCH_2COOH) and ethoxyacetic acid ($CH_3CH_2OCH_2COOH$), the principal metabolites of these two glycol ethers, are responsible for the observed toxic effects, especially those affecting testicles. They are also implicated in the teratogenic effects of the methyl and ethyl ethers of ethylene glycol (as well as the corresponding acetates) which have been described in animals.

The toxicity of the dimethyl ether of ethylene glycol (diglyme) is not well known, but it also seems to have toxic effects on spermatogenesis.

Many nitrogenated solvents (nitroalkanes, nitrobenzene, aniline, etc.) are toxic to blood (producing methemoglobin and hemolysis), while some (2-nitropropane, *o*-toluidine) are carcinogenic in animals.

Dimethylformamide (DMF), an irritant that penetrates skin with ease, is a hepatotoxin whose carcinogenicity is just now being studied. The appearance of testicular cancer in workers exposed to DMF has recently been described.

Hexamethylphosphoric triamide (HMPT), an excellent dipolar aprotic solvent, is a potent carcinogen by inhalation in rats (nasal cavity cancer) at very low concentrations (50 ppb during 2 years in rats).

Preventive measures

As a general rule, the most dangerous solvents must be eliminated and replaced by solvents having similar, though much less toxic, properties. This preventive approach can be applied to the majority of toxic solvents routinely used in laboratories and in the chemical industry.

Thus, within the large family of simple hydrocarbon solvents (alkanes), only hexane possesses any real peripheral neurotoxic properties (polyneuritis). For most of its applications (extraction, chromatography, etc.), hexane can be replaced by other aliphatic alkanes (pentane, methylpentane, heptane) or by cyclic alkanes (cyclohexane, methylcyclopentane, methylcyclohexane).

The replacement of benzene, the only solvent recognized as a carcinogen in human beings (and classified in category 1 by the IARC) should pose no difficulty. The hydrocarbon whose solvolytic properties most closely resemble that of benzene is toluene, though the latter is, like benzene, a central neurotoxin (encephalitis, etc.). In contrast, cyclohexane is devoid of long-term neurotoxic activity and does not seem to possess marked immunotoxic properties. It is sometimes possible to use solvent mixtures to find solvolytic properties similar to benzene (e.g., a mixture of cyclohexane and methyl-cyclohexane).

All chlorinated solvents are neurotoxic and some are hepato or nephrotoxic. Among these solvents, the least toxic appear to be dichloromethane (methylene chloride) and especially 1,1,1-trichloroethane. The use of fluorinated solvents such as 1,1,1-trifluoroethane (fluorocarbon 113) is no longer recommended; these solvents are atmospheric pollutants since, together with other fluorocarbons, they contribute to destruction of the ozone layer. It should be noted that chlorinated solvents such as 1,1,1-trichloroethane also contribute to ozone destruction and their expulsion into the atmosphere should be severely restrained.

The use of methanol as a cleaning solvent is not recommended. This solvent easily penetrates skin, attacking the optic nerve. It can be replaced by ethanol, which is much less neurotoxic. If possible, dilute ethanol is preferable as it is nonflammable.

Ethylene glycol and diethylene glycol are particularly toxic by ingestion (acute anuric tubulopathy, acidosis, convulsions, hepatic cytolysis, etc.) and may be replaced by weakly toxic propylene glycol for certain uses. The methyl and ethyl ethers of ethylene glycol and of diethylene glycol, which all have similar toxic properties (bone marrow, lymph glands, central nervous system, testicles, teratogenicity) may be replaced by propyl or butyl ethers, devoid of these types of toxicity. Similarly, ethers of propylene glycol (1-alkoxypropan-2-ol, $CH_3CHOHCH_2OR$) are relatively nontoxic and may be used in place of the corresponding ethers of ethylene glycol and diethylene glycol.

Because of its potential carcinogenic activity, dioxane must be replaced for many of its multiple uses (scintillation solvents, reaction media, etc.). In organic synthesis, it may be replaced by other ethers such as tetrahydrofuran which does not appear to be toxic.

Nontoxic nitroalkanes (nitromethane, nitroethane, 1-nitropropane) may be used as substitutes for 2-nitropropane, a hepatic carcinogen in rats. Amides such as formamide and acetamide as well as their methylated derivatives are generally hepatotoxic and, in the case of formamides, teratogenic in rats. The best replacement solvent is N-methyl-2-pyrrolidone which is much less toxic.

Dioxane, the formamides, acetonitrile (a similar, though weaker, toxicity as that of cyanides) as well as hexamethylphosphoric triamide (HMPT) possess highly specific solvolytic properties and their replacement by less toxic solvents is often more problematic.

In practice, the use of benzene (a known carcinogen in humans) and of solvents known for their carcinogenic properties in animals (chloroform, carbon tetrachloride, 1,2-dichloroethane, trichloroethylene, tetrachloroethylene, 1,4-dioxane, 2-nitropropane, HMPT, etc.) should be limited as much as possible. The use of benzene as a solvent must be eliminated in laboratories.

Protection against neurotoxic solvents (hexane, toluene, halogenated solvents, etc.) must also be assured, keeping in mind that many of these solvents can potentiate each other's toxic activity (e.g., hexane and methyl ethyl ketone).

Some solvents depress the immune system and weaken the organism's defenses. This may promote various aggressions (microbial, viral, etc.) and diminish resistance to tumor development.

In the case of reputedly toxic solvents (aromatic hydrocarbons, halogenated derivatives, amides, etc.), it is necessary to consider women of child-bearing age or who are nursing as a high-risk population. they should accordingly be effectively protected from toxic environments.

Individual susceptibility to the toxic effects of xenobiotics such as solvents implies that preventive measures require collective action and vigilance. Everyone must feel concerned by these problems and endeavor to improve the working environment. Table 1.34 summarizes the suggested replacements for many of the more toxic solvents. These data can obviously be modified as experimental and epidemiological studies progress and economic situations evolve.

Toxic solvent	Site of action	Observed disorder	Replacement solvent
Hexane	PNS	Polyneuritis	Pentane, Heptane, Cyclohexane, Methylcyclohexane
Benzene[**]	Bone marrow	Aplastic anemia[**] Leukemia[**]	Toluene (+/-) Cyclohexane
Chloroform[*]	CNS, Heart, Liver, Kidneys	Nervous disorders, Myocardia, Hepatonephritis, Cancer (liver, kidneys)	Methylene chloride (?) 1,1,1-Trichloroethane
Carbon tetrachloride[*]	Liver, Kidneys	Cirrhosis, Nephritis	Methylene chloride (?) 1,1,1-Trichloroethane
1,2-Dichloro-ethane[*]	CNS, Liver, Kidneys	Nervous disorders, Hepatonephritis, Adenocarcinoma[*]	Methylene chloride (?)
Trichloro-ethylene[*]	CNS, Heart, Liver, Kidneys	Nervous and heart disorders	1,1,1-Trichloroethane Methylene chloride (?)
Tetrachloro-ethylene[*]	Heart, CNS, Liver, Kidneys	Liver cancer	
Methanol	Optic nerve	Generalized disorders, Blindness	Ethanol

Ethylene glycol	Kidneys	Nephritis, Teratogenic effects	Propylene glycol
Dioxane*	Liver, Kidneys, Nasal cavity	Hepatonephritis, Cancer (nasal cavity, liver...)	Tetrahydrofuran
2-Nitropropane*	Liver	Liver cancer	Nitromethane, 1-Nitropropane
Formamide	Liver	Hepatic disorders	N-Methyl-2-pyrrolidone
N-Methyl-formamide		Testicular cancer	
N,N-Dimethyl-formamide**		Teratogenic effects	
Acetonitrile	CNS	Generalized disorders (convulsions)	
Carbon disulfide	CNS, PNS	Generalized disorders, Polyneuritis, Blindness	
Hexamethyl-phosphoric triamide (HMPT)*	Nasal cavity	Cancers* (nasal cavity)	DMPU, DMEU, TES
Methoxyethanol	CNS	Encephalitis	Propoxyethanol
Ethoxyethanol	Bone marrow, Reproductive system, Teratogenic effects	Anemia, Testicular atrophy	Butoxyethanol, 1-Alkoxy-2-propanol

CNS : Central Nervous System   PNS : Peripheral Nervous System
DMPU : N,N'-Dimethylpropyleneurea
DMEU : 1,3-Dimethyl-2-imidazolidinone
TES : N,N,N',N'-Tetraethylsulfamide
*: Carcinogenic in animals   **: Carcinogenic in man

Table 1.34   Principal toxic solvents and their nontoxic substitutes.

1.3.5 Toxicochemistry and the Prediction of Toxic Risks

1.3.5.1 Definition of "Toxicochemistry"

Toxicochemistry, a new discipline at the interface of chemistry and toxicology, allows a molecular approach to the prediction of a xenobiotic (i.e., foreign to the organism) chemical substance's toxicity. Being a study of the molecular interactions between xenobiotic substances and their biological targets, toxicochemistry requires a multidisciplinary approach (Figure 1.16).

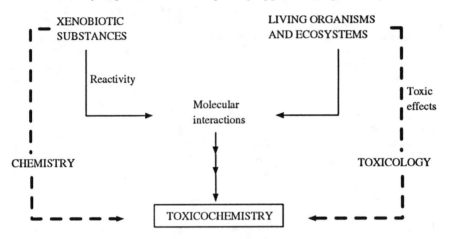

Figure 1.16   Chemical approach to toxicology.

1.3.5.2 Definition of a "Toxicophore"

A relatively simple way of understanding the relationships between the chemical structure of a given molecule and its toxic activity is to look for the presence or absence of a "toxicophore" on the molecule. A toxicophore may be recognized by the fact that any chemical modification which it undergoes modifies or completely suppresses its toxic activity. Similarly, the transfer of a toxicophore to an innocuous molecule is likely to confer toxic activity to the latter. Yet, chemical modifications brought to the molecule at positions other than on the toxicophore should alter only the intensity of the activity but not its type.

For compounds that need to be metabolized by the organism in order to become toxic (i.e., protoxins), a "protoxicophore" may be defined whose enzymatic activation will ultimately yield a toxicophore.

This idea of a toxicophore, just as that of a pharmacophore (the part of a molecule which carries therapeutic activity) must be interpreted with the greatest caution and all excessive generalizations are to be avoided.

As an example, the introduction of a sulfonic acid group onto an arylamine nucleus suppresses all of the latter's genotoxic activity probably because the compound is thereby rendered water-soluble and inaccessible to metabolism by liposoluble enzymes. Thus, 4-aminobiphenyl, a potent bladder carcinogen, becomes nongenotoxic when a sulfonic acid function is introduced in the *para* position. In the case of 4-dimethylaminoazobenzene, however, a mutagenic and hepatocarcinogenic liposoluble dye, the addition of a sulfonic acid moiety (as its sodium salt) to the *para* position yields a water-soluble dye (helianthine or methyl orange, a color indicator) which still possesses mutagenic activity but which is no longer carcinogenic in animals (Table 1.35).

This notion of toxicophores and of protoxicophores allows, for a given chemical series, a rational approach to understanding the relationships between chemical structure and toxicity, particularly long-term toxicity (mutagenic, carcinogenic, teratogenic, immunotoxic, or organotoxic effects). For example, examination of the chemical structures found in Table 1.18 permits identification of several toxicophores responsible for the mutagenicity and/or the carcinogenicity of these molecules. Many direct carcinogens, that is,

STRUCTURE	GENOTOXIC ACTIVITY	
	MUTAGEN (Ames test)	CARCINOGEN (animal experiments)
4-Aminobiphenyl	+	+
4-Aminobiphenylsulfonic acid	-	-
4-Dimethylaminoazobenzene	+	+
Sodium 4-dimethylaminoazobenzenesulfonate (helianthine)	+	-

Table 1.35 Influence of a sulfonic acid function in structures possessing genotoxic activity.

substances that act directly on nuclear DNA without the intervention of any enzymatic system, possess a reactive toxicophore.

Among the toxicophores listed in Table 1.36 may be cited:

PROTOXICOPHORES	TOXICOPHORES
Alkenes	Epoxides
Polyarenes	Diol-epoxides
	$R_1 - X \begin{cases} \text{Cl, Br, I} & \text{Halides} \\ \text{O - SO}_2\text{ - R}_2 & \text{Sulfonates} \\ \text{O - SO}_2\text{ - O - R}_2 & \text{Sulfates} \end{cases}$
	$R - O - CH_2 - Cl$ α-Chloromethyl ether
	Bis(chloroethyl) derivatives
	NR_5 = Aziridines, S = Episulfides
R_1 = H, R... Aromatic amines	H = Hydroxylamines, R_2= Hydroxylamine esters
N = O Nitrosoamines, Hydrazines	
Nitroarenes	Nitro radical anion

Table 1.36 Principal protoxicophores and toxicophores encountered in carcinogenic substances.

-alkyl sulfates, R_1—O—SO_2—O—R_2 (dimethyl sulfate, diethyl sulfate, diisopropyl sulfate, etc.)

-sulfonates, R_1—O—SO_2—R_2 (methyl methanesulfonate, ethyl methanesulfonate, methyl fluorosulfonate, methyl p-toluenesulfonate, etc.)

-α-chloromethyl ether moiety, R—O—CH_2Cl (chloromethyl methyl ether, bis(chloromethyl) ether, etc.)

-bis(chloroethyl) moiety linked to a sulfur atom (yperite) or a nitrogen atom (caryolysine)

-three-atom heterocycles, the heteroatom being oxygen (ethylene oxide, etc.), nitrogen (aziridine, etc.), or sulfur (ethylene sulfide)

All these functions confer potent alkylating properties to the molecules to which they are attached. This allows formation of a stable covalent bond between the foreign molecule and cellular macromolecules having reactive nucleophilic functional groups (DNA, RNA, proteins).

1.3.5.3 Applications of Toxicochemistry: Some Examples

Hexane: a peripheral neurotoxic solvent and its replacement

Among hydrocarbon solvents belonging to the simple alkane series (linear, unbranched hydrocarbons), it is surprising that only hexane (a six-carbon compound) possesses potent neurotoxic properties. This toxicity manifests itself as a progressive limb paralysis (polyneuritis). Long-term intoxication of animals with hexane produces degeneration of nerve fiber terminals, thereby suppressing neural transmission (Figure 1.17).

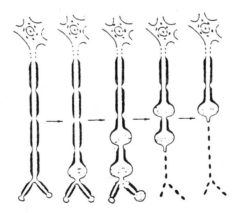

Figure 1.17 Degeneration of a peripheral nerve fiber during the course of hexane-induced polyneuritis.

A simple comparison of the structural formulas of C_5–C_7 linear alkanes does not allow prediction that, of these, only hexane (C_6) is a peripheral neurotoxin.

C_5

H_3C ⌒⌒ CH_3

Pentane

C_6

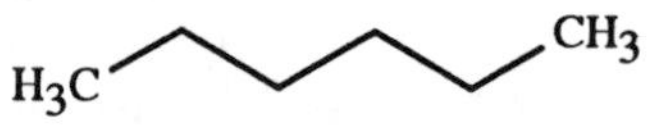

H_3C ⌒⌒⌒ CH_3

Hexane

C_7

H_3C ⌒⌒⌒⌒ CH_3

Heptane

Differences among these compounds must thus exist during the course of their transformation in the organism.

In order to expel highly liposoluble alkanes from the organism, they must be metabolized such that they are transformed into water-soluble compounds (metabolites) that can be eliminated via urine. This biotransformation is classically performed in two successive steps. In the first step, selective oxidation of the next to last carbon atom (position ω-1) transforms the alkane into a secondary alcohol (2-alkanol). To this alcohol, which is still not sufficiently water-soluble to allow its elimination, small, highly polar endogenous molecules such as glucuronic acid (a glucose derivative) or sulfuric acid must then be attached. It is the resulting glucuronides and sulfates (in the form of alkaline salts) which are found in urine and which in some cases serve as biological markers of hexane poisoning (Figure 1.18).

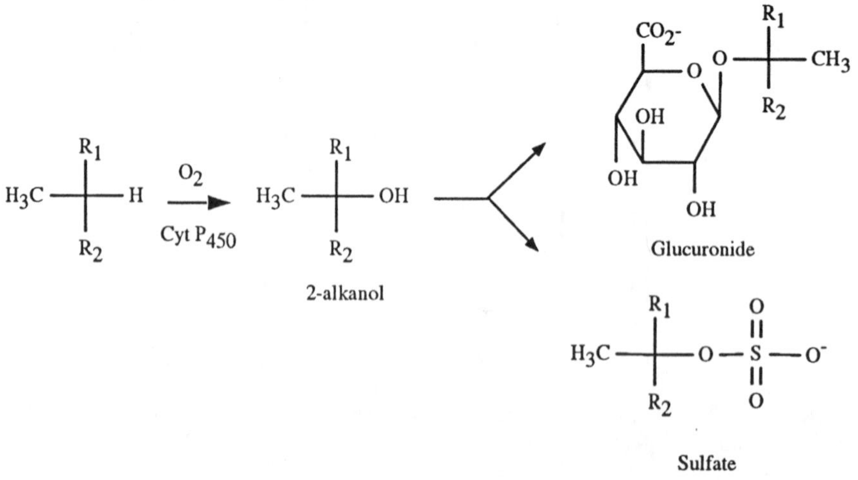

Figure 1.18   Transformation of an alkane into a water-soluble metabolite.

Pentane, heptane, and, to a small extent, hexane, are eliminated in this way. In the case of hexane, the 2-hexanol formed during the first oxidation step is not eliminated by conjugation but, rather, is transformed by various enzymes into methyl butyl ketone and finally into the ultimate metabolite, 2,5-hexanedione, responsible for neurotoxicity (Figure 1.19). In humans, this γ-diketone is the principal metabolite found in blood and urine. It is the causative agent of the polyneuritis that appears after long-term exposure to hexane or methyl butyl ketone.

2,5-Hexanedione can, like other γ-diketones, react by way of a Paal-Knorr reaction with primary amine functions of certain nerve cell constituents (proteins, polyamines, etc.). The pyrrole heterocycle formed as a result may participate in denaturing these vital structures. This hypothesis was confirmed in 1981 by the isolation of neurofilament proteins whose terminal free amine functionalities are blocked in the form of pyrrole derivatives (Figure 1.19).

It is thus evident why pentane (n = 2) and heptane (n = 4), which are not metabolized to 2,5-hexanedione type γ-diketone derivatives, are not peripheral neurotoxins. This difference in toxicity is the result of hexane's transformation into a neurotoxic metabolite containing a γ-diketone (Figure 1.20). The latter

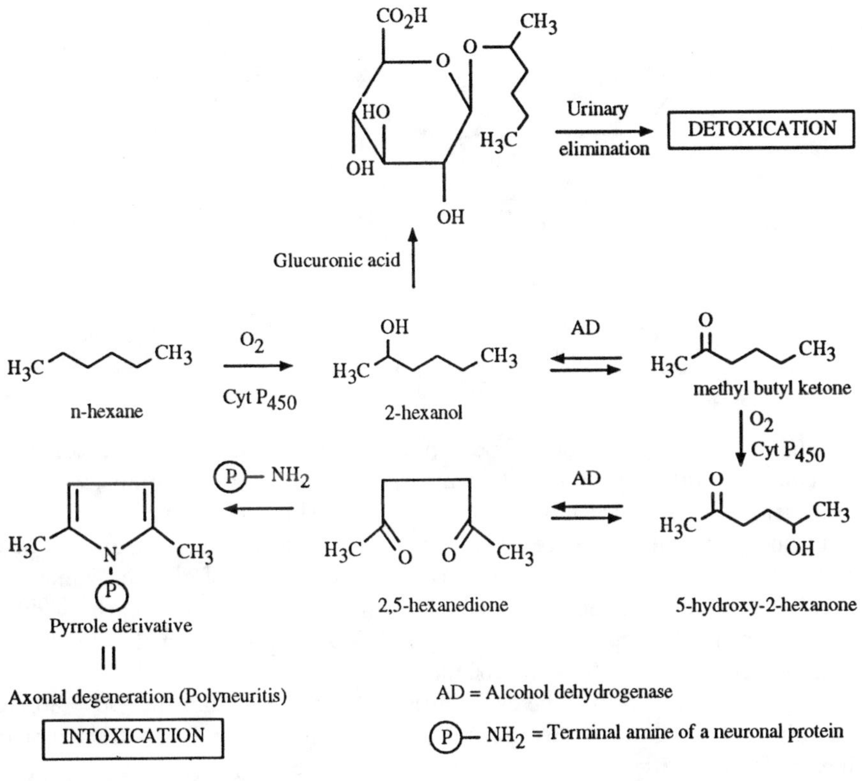

Figure 1.19   The transformation of hexane into 2,5-hexanedione.

may thus may thus be considered to be the toxicophore responsible for the peripheral neurotoxicity of hexane and of the various intermediate metabolites which in the end yield 2,5-hexanedione.

Methyl substituents at the 3,4 positions of 2,5-hexanedione ($R_2=R_4=CH_3$) increase the peripheral neurotoxic potency of the γ-diketone (20–30 times) by facilitating formation of the pyrrole ring. On the other hand, disubstitution with methyl groups at position 3 ($R_2=R_3=CH_3$) prevents cyclization to a pyrrole ring; 3,3-dimethyl-2,5-hexanedione is thus free of peripheral neurotoxic properties.

$$\left.\begin{array}{l} R_1 \\ R_6 \end{array}\right\} CH_3$$

$$\left.\begin{array}{l} R_2 \\ R_3 \\ R_4 \\ R_5 \end{array}\right\} H, CH_3$$

γ-DIKETONE " TOXICOPHORE "

3,4-dimethyl-2,5-hexanedione :
a potent peripheral neurotoxin

3,3-dimethyl-2,5-hexanedione :
not a peripheral neurotoxin

The dissolving properties of pentane, hexane, and heptane being, for most practical laboratory purposes, equivalent, only pentane and heptane, which, in contrast to hexane, are not peripheral neurotoxins, should be employed. In choosing a solvent to replace hexane, physicochemical properties, and particularly volatility, must be taken into account. In the case of alkanes, which are highly flammable hydrocarbons, it is preferable to replace hexane (bp = 69°C at 760 mm Hg) by heptane (bp = 98.5°C at 760 mm Hg) rather than by highly volatile pentane (bp = 36°C at 760 mm Hg).

It should be noted that the cyclic form of hexane, cyclohexane (bp = 80.7°C at 760 mm Hg), is devoid of peripheral neurotoxicity and can thus replace hexane for most uses despite its higher cost (Table 1.37).

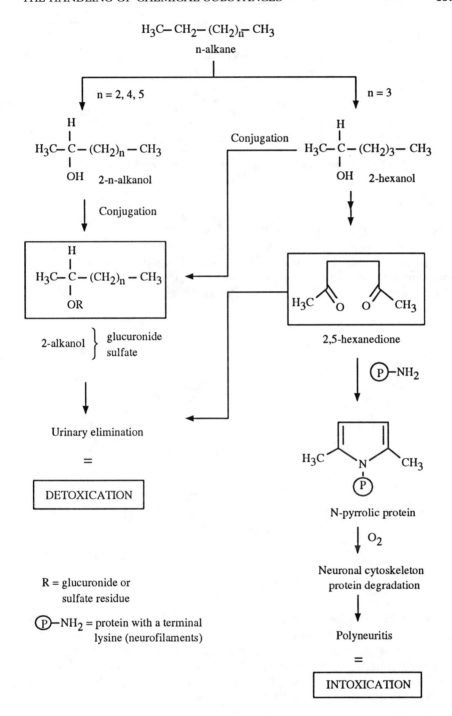

Figure 1.20 Different metabolic routes for alkanes.

SOLVENT	BOILING POINT (at 760 mm Hg)	PERIPHERAL NEUROTOXICITY
Pentane	36°C	-
Hexane	69°C	+
Cyclohexane	80.7°C	-
Heptane	98.5°C	-

Table 1.37   Criteria in choosing a replacement solvent for hexane.

Nephrotoxic alkanes

Experiments on male rats have shown that exposure to mixtures of saturated hydrocarbons containing branched-chain alkanes (isoparaffins) affects kidneys in a particular way (degeneration of the tubuli contorti accompanied by hyaline protein deposits). Isooctane (2,2,4-trimethylpentane) and various cyclic alkanes such as decalin also affect male rat kidneys in an identical manner.

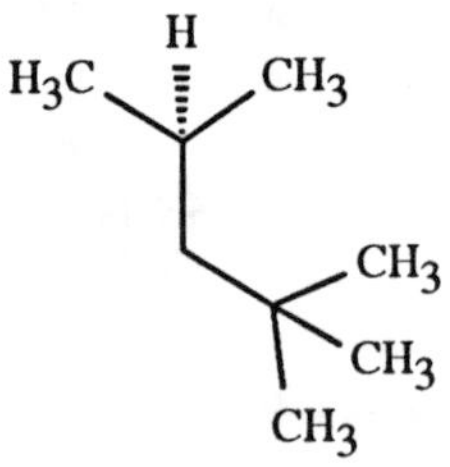

2,2,4-Trimethylpentane

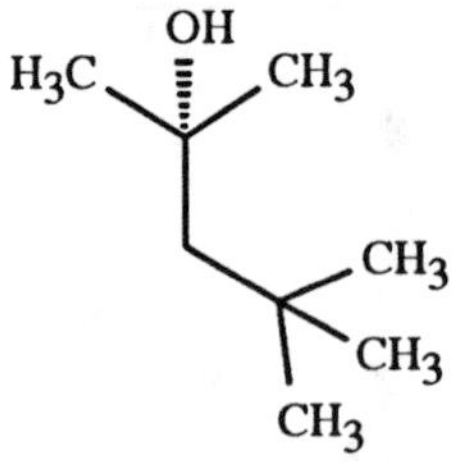

2,2,4-Trimethyl-2-pentanol

cis

trans

Decalin

No correlation between alkane structures and their nephrotoxic properties observed in male rats (but not in female rats) has yet been found. It is possible that toxicity is the result of metabolic routes that lead to the accumulation of toxic metabolites in the renal tubules. These metabolites then associate with a renal protein (α-2μ-globulin) specific to male rats. In the case of isooctane, the principal metabolite is 2,2,4-trimethyl-2-pentanol which also forms a complex with α2μ-globulin.

In workers exposed to petroleum products, renal disorders (glomerulonephritis, cancers, etc.) have been described, though no correlation has yet been established between the nature of these compounds and their eventual nephrotoxicity. The example of nephrotoxic alkanes demonstrates how difficult it is to establish simple relationships between chemical structure and a given type of toxicity. In animal experiments, the species, sex, nutritional factors, and so on sometimes have a determinant role in the biotransformation of a xenobiotic. Because of this, all extrapolations are subject to caution, even within analogous series.

Replacement compounds for benzidine, a bladder carcinogen

Benzidine is one of the best known compounds of the aromatic amine family, constituting the basic starting material for the synthesis of azo dyes (aniline dyes). Thus, benzidine is employed in the manufacture of more than 250 dyes, though this use is now subject to severe regulations in many countries (e.g, United States, France) and its industrial applications are tending to diminish.

Benzidine is mainly used in laboratories for detection, for example, in chemistry (detection of peroxides, chlorine, nitriles, sulfates, carbohydrates, etc.), in biochemistry (detection of peroxidases, etc.) and clinically (detection of blood traces, etc.). Benzidine and its derivatives (*o*-dianisidine, *o*-toluidine) are even used for detection in student laboratory courses.

Benzidine, however, is one of the most dangerous human carcinogens found in laboratories. As with other aromatic amines (2-naphthylamine, 4-aminobiphenyl, *o*-toluidine, etc.), benzidine can lead over extremely variable latency periods (between a few months to 15–20 years or more) to the appearance of particularly serious cases of bladder cancer. Its mechanism of action is still not well known, and a variety of mechanisms appear to be implicated depending on the site where it is metabolized. In the bladder, enzymatic systems having peroxidase activity (peroxidases or cyclooxygenases of arachidonic acid metabolism) oxidize one of benzidine's nitrogen atoms by a radical-type mechanism (single electron oxidation). The first cation radical-type intermediate is then oxidized to form reactive entities (diimines, nitrenium, etc.) which are suspected to be responsible for stable covalent bond formation with macromolecules (DNA, etc.) found in bladder epithelial cells. Carcinogenesis is thereby initiated (Figure 1.21).

It might be expected that by sufficiently encumbering access to benzidine's two nitrogen atoms, the enzymes involved in their oxidation would be prevented from acting, thus precluding formation of the reactive intermediates responsible for toxicity. In fact, the addition of only one substituent to each of the aromatic rings (R_1 and R_3=CH_3, OCH_3, Cl, NH_2, etc.) does not suppress genotoxic activity, the resulting compounds still being mutagenic and carcinogenic. The addition of two substituents to each side of the nitrogen atoms, however, (as in 3,3',5,5'-tetramethylbenzidine or TMB) suppresses mutagenic and carcinogenic activity while maintaining the capacity to detect peroxidases or substances having peroxidase-type activity (cytochrome P_{450}) (Table 1.38).

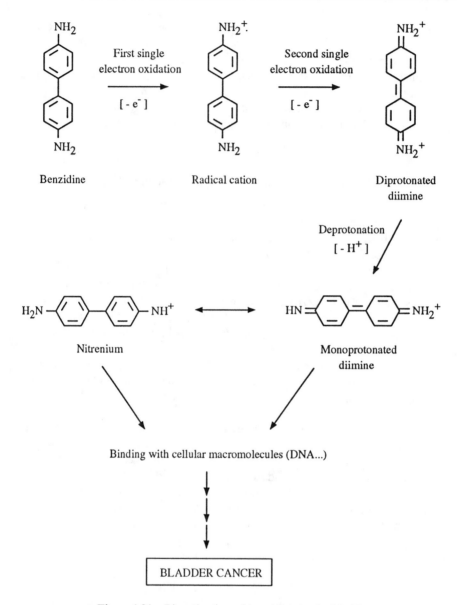

Figure 1.21   Bioactivation of benzidine in the bladder.

In order to verify that steric hindrance around the nitrogen atoms was sufficient to suppress genotoxic activity, *ortho-ortho'* tetrahalogenated derivatives of benzidine were synthesized and their mutagenic activities were tested. While the tetrachloro as well as the tetrabromo derivatives are, like the tetramethyl derivative, inactive, the tetrafluoro derivative demonstrates mutagenic activity equivalent to that of benzidine in certain Ames tests (Table 1.39).

COMPOUND	MUTAGENIC ACTIVITY	CARCINOGENIC ACTIVITY	PEROXIDASE DETECTION (Sensitivity)
H_2N—⬡—⬡—NH_2 Benzidine	+	++	++
H_3C ⬡ CH_3 H_2N—⬡—⬡—NH_2 o-Toluidine	+	++	+++
H_3C ⬡ CH_3 H_2N—⬡—⬡—NH_2 H_3C CH_3 3,3',5,5'-Tetramethylbenzidine	-	-	+++

Table 1.38 Comparison between the genotoxic activity and peroxidase detection sensitivity of various benzidine derivatives.

If the van der Waals radii of these different atoms are considered, it is apparent that only those atoms having a volume greater than that of hydrogen (Cl, CH_3, Br) are able to inhibit this mutagenic activity. Furthermore, 3,3',5-trichlorobenzidine, which has one free *ortho* position, possesses mutagenic activity greater than that of benzidine itself. The presence of halogen atoms such as chlorine and bromine in the *ortho* and *ortho'* positions (positions 3,5 and 3',5') considerably increases the mutagenic activity of benzidine on condition that a nonbulky substituent is present in one of the ortho positions (Z = H or F in the case of 3,3',5,5'-tetrafluorobenzidine in Figure 1.22) (Table 1.39).

It thus appears that, in order to observe mutagenic activity in the benzidine series, it is necessary that at least one of the *ortho* positions be unhindered (Z = H or F). The absence of steric hindrance in this position allows adduct formation between a guanine in the C-8 position of DNA and a reactive intermediate (monoprotonated diimine or nitrenium) resulting from activation of one of benzidine's nitrogen atoms (Figure 1.22).

The presence of mutagenic activity in the benzidine series requires:

–at least one unhindered amine functionality
–at least one free *ortho* position if bulky substituents are present.

As mentioned previously, premature generalizations in the toxicochemistry field should be avoided. In the 4-aminobiphenyl and aniline series, for example, steric hindrance of the nitrogen atom by *ortho* and *ortho'* methyl sub-

Van der Waals radius (Å)	COMPOUND	MUTAGENIC ACTIVITY (Ames test with activation)
H = 1.2	H_2N — benzidine ring with H substituents — NH_2 Benzidine	++
F = 1.35	H_2N — ring with F substituents — NH_2 3,3',5,5'-Tetrafluorobenzidine	++
Cl = 1.80	H_2N — ring with Cl, Cl, Cl, H substituents — NH_2 3,3',5 -Trichlorobenzidine	+++
Cl = 1.80	H_2N — ring with Cl, Cl, Cl, Cl substituents — NH_2 3,3',5,5'-Tetrachlorobenzidine	0
CH_3 = 2.0	H_2N — ring with H_3C, CH_3, H_3C, CH_3 substituents — NH_2 3,3',5,5'-Tetramethylbenzidine	0

Table 1.39   Mutagenic activities in the benzidine series.

stituents does not suppress genotoxic activity. This may be due to the possibility of metabolism occurring at the position *para* to the amine, which is free, in contrast to that in the benzidine derivatives (Table 1.40).

In general, 3,3',5,5'-tetramethylbenzidine (TMB) can be used in place of benzidine for peroxidase detection (cytochemistry, cytoimmunochemistry, clinical, etc.). In contrast to benzidine and to *ortho* disubstituted benzidines ($R_1=R_2=CH_3$, OCH_3, Cl, Br, NH_2, etc.) 3,3',5,5'-tetramethylbenzidine is not

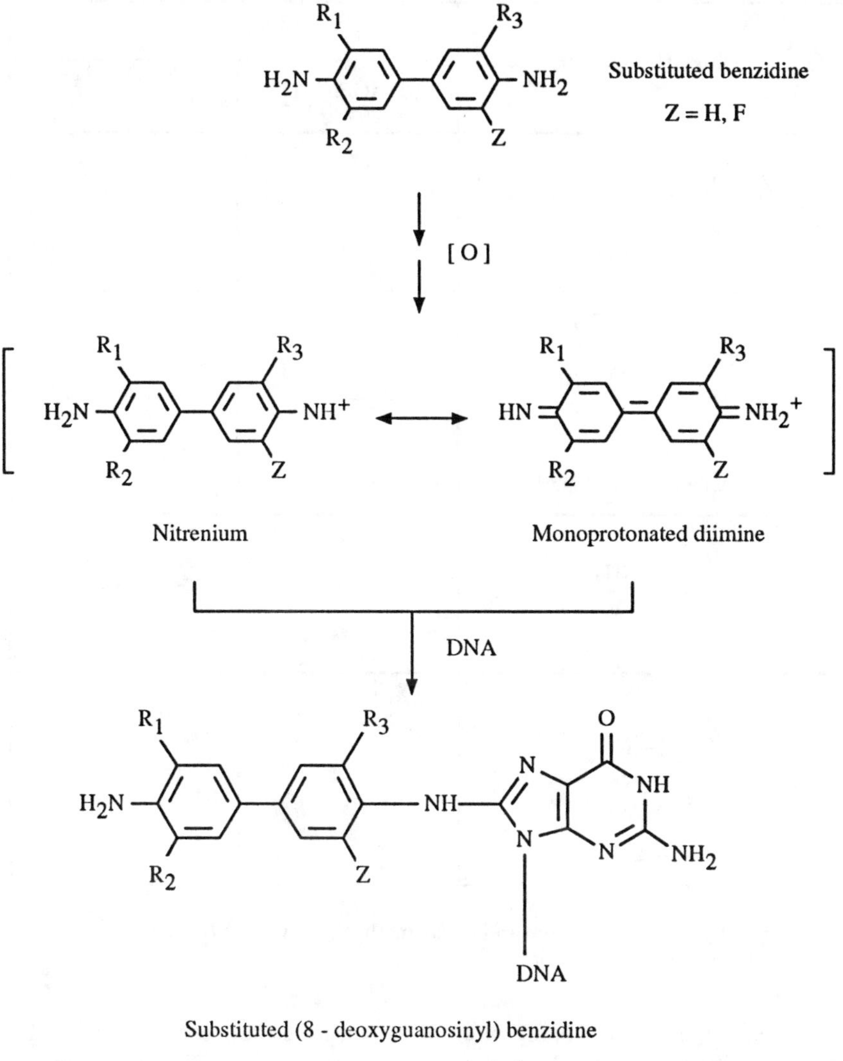

Figure 1.22 Adduct formation between guanine of DNA and *ortho*-substituted benzidines.

sensitive to polymerization (a process which requires a free *ortho* position), thereby excluding its use for certain characterizations such as the detection of hemoglobin on acidic polyacrylamide gels. 3,3'-Diaminobenzidine, however, widely used in cytochemistry, is an excellent substitute for benzidine, though it is mutagenic and weakly carcinogenic in animals. For many users, a substitution product must have qualities that are equivalent to those of the substance it is meant to replace. Unfortunately, as the benzidine example illustrates, this is not always possible in cases where substances having multiple applications must be replaced. Thus, for the majority of uses, 3,3',5,5'-tetramethylbenzidine

COMPOUND	GENOTOXIN	
	MUTAGENICITY	EXPERIMENTAL CARCINOGENESIS
4-Aminobiphenyl	+	+
3,5-Dimethyl-4-aminobiphenyl	+	?
Aniline	-	+
2,6-Dimethylaniline	?	-

Table 1.40 The influence of *ortho* and *ortho'* methyl substituents on genotoxicity in the 4-aminobiphenyl and aniline series.

is an excellent substitute, being often more sensitive than benzidine itself. For certain specific techniques, however, only 3,3'-diaminobenzidine gives satisfactory results. All the rigorous methodology necessary in the handling of hazardous substances must then be employed if this compound is used.

Still much research will be necessary to find sufficiently sensitive and nontoxic replacement substances that will guarantee adequate health protection of personnel obliged to work over long periods with these substances.

1.3.5.4 *What Can Be Expected from Toxicochemistry?*

Chemical risk represents an important sector of safety and hygiene and, within this domain, risks associated with the toxic effects of foreign chemical substances (xenobiotics) are by far those which pose the most problems. After all, what else is toxicity but the result of the detrimental action of a xenobiotic substance on biological targets or ecosystems (Figure 1.16)? This is nothing more than molecular interaction.

Toxicochemistry, an interdisciplinary field, can help to bring a molecular explanation (described by chemistry) to an observed toxic effect (described by toxicology) (Figure 1.16). The most important feature of this theoretical approach is its predictive value. Based on sound knowledge of the bioavailability and the reactivity of the ultimate toxin, it is possible in some cases to find within a given series of compounds that or those substances which, all else being equal, will have the least detrimental effects on health, especially over a long period.

Thus, why is it that, within the alkane family, only hexane is a peripheral neurotoxin? The investigation of its biotransformation (Figure 1.19) showed that hexane is mainly metabolized into 2,5-hexanedione which accumulates in blood and nerve tissue, producing the observed toxic effects (degeneration of peripheral nerves). This ultimate metabolite possesses a 1,4-diketone moiety in its structure. It is this moiety that is responsible for the high reactivity and, consequently, the toxicity of the molecule. In contrast, the other alkanes are not transformed into metabolites comprising such a toxicophore.

In this manner, the identification of a toxicophore in a series of compounds having similar properties allows elimination of the toxic compound (e.g., hexane), leaving the possibility of working with those compounds that present the least risk to health (e.g., heptane, pentane, etc.). It is possible to establish relationships between the chemical structure of a given compound and its observed toxicity if, on one hand, details of its bioavailability and its metabolic fate are known and, on the other hand, information is available concerning the reactivity of the metabolites (final or intermediate) implicated in this toxicity. Even within an analogous series, extrapolation to other compounds must always be done prudently and the results only considered as being indicative. Only animal experiments (in at least two species, one being a rodent) and, especially, epidemiology can confirm whether or not a compound, predicted to be toxic by the methods of toxicochemistry, is, in fact, so.

Historically, the study of quantitative structure-activity relationships (or QSAR) was mainly developed by C. Hansch* who took both the thermodynamic and electronic properties of molecules into consideration. In 1981, J.J. Kaufman[†] proposed a computerized method for the elucidation of structure-toxicity relationships with the purpose of an eventual prediction of toxicity. The use of the modeling capabilities of the DARC system, developed by J.E. Dubois,[‡] is well adapted to this type of approach. In theory, QSAR

---

*C. Hansch and A. J. Leo. *Substituent constant for correlation analysis in chemistry and biology.* John Wiley and Sons, New York (1979).

†J. J. Kaufman. Strategy for computer generated and quantum chemical prediction of toxicity and toxicology. *Int. J. Quantum Chem.*, 1981, QBS8, 419–439.

‡J. E. Dubois. *DARC system in chemistry for chemical representation and manipulation of chemical information.* W. T. Wipke, S. Heller, R. Feldmann, and E. Hyde, Eds. John Wiley and Sons, New York (1973).

allows prediction of the nature of the toxicity-carrying toxicophore groups in a given series of molecules. It is important to keep in mind that, in addition to the toxicophore(s) (or protoxicophores which require enzymatic bioactivation), functional groups or other parts of the molecule may play an essential role in determining the bioavailability of a molecule. This is particularly true in the case of detoxifying metabolism, in which easily eliminated water-soluble metabolites may be formed. For example, in the bioactivation of polyaromatic hydrocarbons (PAH), which, like benzo(a)pyrene (BaP), are serious environmental pollutants, many factors must be considered before genotoxicity can be predicted. As has been demonstrated by D.M. Jerina and R.E. Lehr,* the presence of a "bay" region having a highly electrophilic benzylic carbon atom is essential for genotoxic activity, metabolic routes leading to the ultimate metabolites (epoxide-diols) intervening at these sites.

If the mutagenic and carcinogenic activities of such compounds are to be distinguished, however, two other regions, K and L (defined by A. and B. Pullman), must be taken into consideration. The K region appears to be more important in bacterial mutagenesis while the L region is probably implicated in the detoxifying process of compounds having such a nucleus.

Based on the data accumulated by the National Toxicology Program (NTP), J. Ashby and R.W. Tennant[†] conducted a systematic study of the possible correlations between chemical structure, mutagenic potency (Ames tests), and carcinogenic activity in rats. The 301 compounds tested allowed the characterization of several potentially genotoxic DNA-reactive functional groups (or "structural alerts to DNA reactivity"). The presence of such a genotoxicity-carrying group on a molecule is an important element in predicting possible mutagenic and/or carcinogenic activities.

Since short-term mutagenicity tests are not always sufficiently precise to allow prediction of carcinogenic activity in animals, systems making use of artificial intelligence have been devised for this purpose. Among these, the

---

*R. E. Lehr and D. M. Jerina. Metabolic activations for polycyclic hydrocarbons: Structure-activity relationships. *Arch. Toxicol.* 1977, **39**, 1.

†J. Ashby and R. W. Tennant. *Mutation Res.* 1991, **257**, 229–306.

CASE (Computer Automated Structure Evaluation Method) system proposed by H.S. Rosenkranz and G. Klopman* utilizes a molecular graphics approach for defining "active (or genotoxic) fragments" and "inactive fragments" in a given structure. The fragments carrying genotoxic activity are referred to as *biophores* while the inactive fragments are termed *biophobes*.

As an example, the CASE system allowed the moderate mutagenic activity of 1-chloro-2,4-dinitrobenzene to be predicted, the aromatic nitro group being the active biophore. Such nitroarene groups are in fact present in various

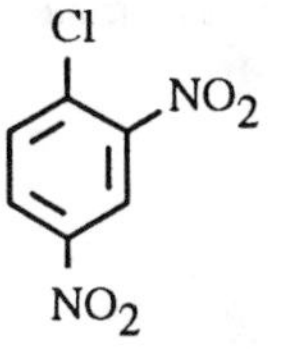

1-Chloro-2,4-dinitrobenzene 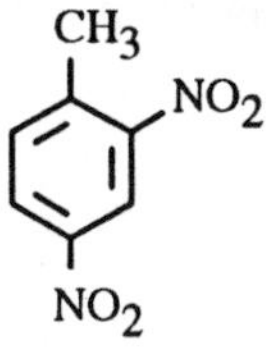   2,4-Dinitrotoluene

genotoxic molecules such as 2,4-dinitrotoluene.

The nitro function also appears to be a good biophore when present on an aliphatic chain. For example, 2-nitropropane (a hepatocarcinogen) and tetranitromethane (a pulmonary carcinogen) are highly genotoxic in animals.

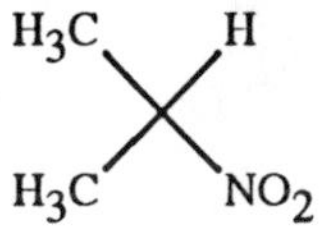

2-Nitropropane 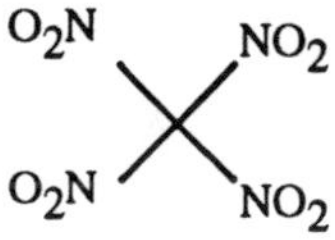   Tetranitromethane

These compounds must be handled using all possible precautions.

1.4 Conclusion

A predictive approach associating information concerning the physicochemical properties of molecules and their interactions with biological targets or ecosystems can allow replacement of toxic substances by compounds presenting fewer risks to users. In order to correctly replace certain compounds used in research laboratories, it is generally necessary to take into account not only the different types of toxicity (organotoxicity, immunotoxicity, genotoxicity, etc.) but also physicochemical criteria (volatility,

---

*H. S. Rosenkranz, C. Mitchell, and G. Klopman, *Mutation Res.* 1985, **150**, 1–11.

flammability, instability, etc.) in order to avoid replacing one risk with another.

In the future, the use of computer technology will no doubt help in the prevention of chemical risks. Artificial intelligence, for example, should allow a better comprehension of the relationships which exist between chemical structure and toxic activity.

Regardless of the approach employed, it cannot be expected that a magical solution will be found that will allow elimination from the workplace of all substances dangerous to human beings or their environment. It is thus necessary to ensure the most effective protection against chemical risks by applying stringent individual and collective preventive measures at all times.

2

Neutralization and Destruction of Chemical Substances

2.1 The Various Types of Wastes

Dealing properly with wastes rests on the two basic principles of categorizing and choice of treatment, the first determining the quality of the second. The better the categorization of different types of wastes, the more effective and economical will be their subsequent treatment (recycling, incineration, neutralization, etc.).

The proper packaging of wastes destined for destruction depends on the type of treatment anticipated and on their physicochemical characteristics (in order to avoid possible incompatibilities).

2.1.1 Common Wastes

Common wastes include all those which may be termed household refuse. These should contain neither chemical, toxic, flammable or reactive substances nor biologically active products. Animal carcasses containing these cited substances must not be eliminated by this route. Common wastes are either incinerated or disposed of in controlled dumping areas. It may easily be understood that the inclusion of chemical, flammable, toxic, or biologically active substances in common waste presents serious dangers both for personnel responsible for their destruction as well as for the environment.

2.1.2 Special Wastes

Special wastes refer to all wastes resulting from industrial or laboratory operations. These may be classified into five categories:

1. Biologically active waste
2. Radioactive waste
3. Gases
4. Used oils and lubricants
5. Chemical waste

Mention should also be made of very particular wastes such as polychlorinated biphenyls (PCBs), dioxins and dioxin precursors such as pentachlorophenol which are subject to special regulations and wastes containing mercury, such as thermometers and mercury batteries. The latter can be distinguished from other types of batteries by the code letters M or N, characteristic of electrochemical systems containing mercury, as well as by the following terms: *Hg* (the chemical symbol for mercury), *mercure* (French), *merkur* or *quick* or *quecksilber* (German), or *quicksilver* (English). These batteries must be recovered and sent to local or national organizations responsible for their recycling.

Use of the environment or of the water system for the disposal of wastes must be limited to substances displaying no toxicity. It is recommended that, in general, nothing should be disposed of in laboratory sinks. Even if a substance is only weakly toxic, chemical reactions can occur in water solution. For example, because of water chlorination, phenol often leads to the forma-tion of chlorophenols, which are both foul-smelling and very toxic to the aquatic environment. Moreover, drain disposal of even reputedly nontoxic substances may have highly undesirable consequences if the drainage network is connected to a biological treatment station since the microorganisms responsible for biodegradation (e.g., of heavy metals) may be destroyed.

2.1.2.1 Biologically Active Wastes

All biologically active substances must be neutralized with certainty before packaging or disposal. In the case of microorganisms, destruction may be achieved either by autoclaving or with chemicals. The following pages list neutralization and destruction techniques, especially those concerning mutagenic and carcinogenic substances.

Animal cadavers must be disposed of separately. The appropriate treatment is incineration, either on-site or in an authorized center that can provide the producer with a certificate of destruction.

2.1.2.2 Radioactive Wastes

Radioactive wastes from nonsealed sources may take the form of solids (paper, plastics, etc.), aqueous or organic solutions, or again putrescible matter (animal cadavers, cell cultures, etc.). Regardless of the nature of the radioactivity, risks associated with the physicochemical form of the radioactive waste should not be ignored.

The disposal of radioactive wastes with ordinary waste is limited by the activity of the material (2 Ci/kg or 74 kilobecquerels/kg) with a daily maximum that depends on the radiotoxicity category. Fractionated disposal or the storage of radioelements of half-life less than 100 days allows reduction of the quantity of radioactive material that must be disposed of at one time.

The packaging of radioactive wastes is severely regulated (depending on their physicochemical form and activities) and in all cases it is advantageous to group the wastes to obtain the best treatment conditions and removal fees.

2.1.2.3 Gases

Compressed gas cylinders are subject to the regulations concerning equipment operating under pressure. These regulations render obligatory the testing of cylinders every 2 years in the case of corrosive gases and every 5 years for noncorrosive gases. Application of these rules allows the problems associated with the long-term storage of unused cylinders to be avoided (e.g., concerns about the contents, frozen valves, etc.).

Once a cylinder is considered waste, it must be returned to its owner who is responsible for it. If the laboratory owns the cylinder, it must be emptied (see the following paragraph), purged, rinsed, filled with water, and punctured in order to render it unusable. This procedure must be employed only in the case of nontoxic and noncorrosive gases. In all other cases, the distributor or a specialist should be consulted.

Toxic or corrosive gas leaks expected in the course of operations must be neutralized by bubbling or filtering.

2.1.2.4 Used Oils and Lubricants

Used oil products must be treated separately from other chemical wastes (solvents, aqueous solutions, etc.). The separation of black oils, clear oils, soluble oils, and cutting oils is recommended. Soluble mineral and organic oils must also be kept separately. Note that some states treat waste oil as hazardous waste. Check with state agencies to ensure that proper storage and handling/shipping requirements are met.

2.1.2.5 Chemical Wastes

Two categories of chemical wastes must be considered:

- Solvents and acids which can be treated in larges volumes (30–200 l) and which may be considered homogeneous
- Residues from reactions or substances stored in quantities of less than

10 kg which can be removed separately or neutralized in the laboratory. (Note that neutralization may be considered treatment and fall under the Treatment, Storage, and Disposal [TSD] hazardous waste regulations. Check with state agencies.)

Large volumes

In the case of large volumes, initial treatment consists in sorting out the different substances to obtain lots as homogeneous as possible. Thus, halogenated solvents should be separated from nonhalogenated solvents, because their destruction requires different techniques.

There exist companies that offer to buy used solvents for purification and recommercialization. In dealing with such firms, vigilance with regard to their competence must be maintained and a perfect knowledge of the composition of the mixtures provided is essential. The smaller the volume to be removed, the higher the cost of its removal. Thus, when the quantities of waste solvents produced are large enough (at least about 2,000 l per year), their recovery by purification may be justified. If the solvent is a hazardous waste, the transporter must have a hazardous waste transporter's license.

The proper management of stocks of both new products and of wastes (storage area, packaging, labeling, duration of storage) is the best means of assuring safety and of saving time and money.

Reaction residues or wastes in small quantities

The treatment of residues and wastes resulting from laboratory manipulations may be effected in the workplace by the workers themselves. This is the preferred procedure since problems of sorting, packaging, and labeling of these wastes together with administrative and financial constraints associated with their removal by exterior firms are avoided. The techniques for the neutralization of a large number of chemical families and substances will be described later.

These techniques of neutralization or destruction must be used only on small quantities, by competent workers and with the necessary equipment. Certain substances are very reactive or very toxic. The application of these techniques under improper conditions can be extremely dangerous.

If the producers of waste decide not to destroy the substances themselves, they must assure the material is removed under the best possible conditions, taking particular care in packaging the materials and in informing the persons responsible for their removal of the risks involved.

With respect to packaging, glass containers should be avoided. The contents should be precisely identified and in such a way that no confusion is possible (particular attention should be paid to packaging which includes the name of the contents or which suggests the presence of a particular substance by its shape or composition).

2.2 Destruction of Alkaline Acetylides and Cuprous Acetylides

Examples

Potassium acetylide	$H - C \equiv C - K$
Sodium acetylide	$H - C \equiv C - Na$
Cuprous acetylide	$H - C \equiv C - Cu$

Principle of destruction

Decomposition of the metallic acetylide in the presence of hydrochloric acid.

Example: decomposition of sodium acetylide

$$H - C \equiv C - Na \ + HCl \longrightarrow \ H - C \equiv C - H \ + NaCl$$

In ethynylation reactions (the addition of sodium acetylide to a carbonyl compound), excess alkaline acetylide can be destroyed by the addition of excess ammonium chloride to the reaction mixture.

Special precautions

Protective goggles
Rubber gloves
Safety shield
Work under a well-ventilated fume-hood

Destruction

To a solution of the cooled acetylide, carefully add, while stirring, an excess of 1 N HCl solution. Allow 15 minutes for reaction. Dilute with water and store for waste disposal.

2.3 Destruction of Calcium Acetylide (Calcium Carbide)

Calcium acetylide, commonly called calcium carbide (CaC_2), is a stable grayish-white solid that reacts with water to liberate acetylene.

$$CaC_2 + 2 H_2O \longrightarrow Ca(OH)_2 + \ H - C \equiv C - H$$

In the presence of a minimal amount of water, calcium carbide liberates acetylene which, in the presence of air, may lead to an explosive mixture capable of igniting due to the exothermic nature of the reaction.

Calcium carbide generally contains calcium phosphide as an impurity which, in the presence of water, liberates phosphine, a highly toxic and flammable gas.

Principle of destruction

Calcium carbide may be destroyed by the action of 6 N aqueous HCl solution.

$$CaC_2 + 2\,HCl \longrightarrow CaCl_2 + \quad H-C\equiv C-H$$

Special precautions

Protective goggles
Rubber gloves
Work under a well-ventilated fume-hood

Elimination of spills

   −Cover with dry vermiculite
   −Carefully collect the spill
   −Place in a plastic bag
   −Destroy as noted under Destruction.

Destruction

In a 2 l, three-necked, round-bottom flask equipped with an efficient stirrer, a dropping funnel, a thermometer, and a nitrogen inlet and outlet, add:

   −600 ml of dry toluene and, while stirring,
   −50 g of calcium carbide

Cool the suspension in an ice bath under a stream of dry nitrogen. Cautiously add over 5 hours while monitoring the temperature:

   −300 ml of 6 N HCl

After completion of the addition, allow the mixture to stir 1 hour, then pour it into a separatory funnel. The organic phase is eliminated along with other nonhalogenated solvents while the aqueous phase is stored separately for waste disposal.

Remarks

Phosphides such as calcium phosphide (Ca_3P_2), aluminum phosphide (AlP), and so on also react with water liberating phosphine (PH_3), a highly toxic gas (pulmonary aggression) which may also ignite spontaneously in air due to the presence of traces of diphosphane (H_2PPH_2).

$$AlP + 3\,H_2O \longrightarrow Al(OH)_3 + PH_3 \nearrow$$

Their destruction must be effected with great caution, in a fume-hood, by adding them to cold water. If a dilute HCl solution is used, an explosion may occur.

2.4 Destruction of Hydrogen Cyanide and Cyanides

Examples

Hydrogen cyanide	HCN
Sodium cyanide	NaCN
Potassium cyanide	KCN
Calcium cyanide	$Ca(CN)_2$
Copper (I) cyanide	CuCN

Principle of destruction

Solutions containing cyanides must never be acidified, otherwise highly toxic hydrogen cyanide will be liberated.

$$K - C \equiv N + HCl \longrightarrow H - C \equiv N + KCl$$

Highly basic products such as sodium hydroxide or potassium hydroxide or even alkaline cyanides must never be suddenly added to liquid hydrogen cyanide, which can polymerize rapidly and explode.

Concentrated solutions of hydrogen cyanide (20% in water or 40% in ethanol) can easily be destroyed by combustion. For the destruction of hydrogen cyanide itself, the methods described for the destruction of cyanides may be used as long as the solution is made strongly alkaline as a precaution (pH 10). Several techniques may be used to destroy cyanides.

Complexation in the form of ferrocyanides

An alkaline solution of cyanide treated with an excess of ferrous sulfate gives ferrous ferrocyanide which can be separated by filtration or else transformed into ferric ferrocyanide (Prussian blue) by oxidation in the presence of a ferric salt (perchlorate or sulfate).

$$6\,NaCN + FeSO_4 \longrightarrow Na_4Fe(CN)_6 + Na_2SO_4$$

Sodium ferrocyanide

$$Na_4Fe(CN)_6 + 2\,FeSO_4 \longrightarrow Fe_2Fe(CN)_6 + 2\,Na_2SO_4$$

Ferrous ferrocyanide

$$3\,Fe_2Fe(CN)_6 + Fe_2(SO_4)_3 \longrightarrow Fe_4Fe(CN)_6 + FeSO_4$$

Prussian blue

Oxidation by sodium hypochlorite

Cyanides are rapidly oxidized by sodium hypochlorite with formation of cyanates:

$$Na - C \equiv N + NaOCl \longrightarrow Na - O - C \equiv N + NaCl$$

The oxidation is a two-step process:

–The first step consists of the formation of cyanogen chloride:

$$NaCN + NaOCl + H_2O \longrightarrow ClCN + 2\,NaOH$$

–Cyanogen chloride, highly toxic, is then destroyed in the strongly alkaline medium (pH 9.5):

$$ClCN + NaOH \longrightarrow Na^+\,{}^-OCN + NaCl + H_2O$$

–The resulting cyanates, which are relatively nontoxic, can finally be degraded into carbon dioxide and nitrogen by acidification of the solution followed by treatment with excess sodium hypochlorite:

$$2\,Na^+\,{}^-OCN + 3\,NaOCl + 2\,HCl \longrightarrow 2\,CO_2 + N_2 + 5\,NaCl + H_2O$$

In the course of this oxidation of cyanide to cyanate, it is necessary to work under highly basic conditions (pH 12) in order to avoid formation of cyanogen chloride.

This oxidation being very exothermic, the cyanide concentration must not exceed 1–$2\,g/l$ of CN^- in order to avoid release of cyanogen chloride as a result of a temperature elevation.

This technique can be applied to water-soluble cyanides (NaCN, KCN, etc.) and to ferrocyanides $Fe(CN)_6K_4$ and ferricyanides $Fe(CN)_6K_3$.

In calculating the quantity of sodium hypochlorite necessary to oxidize a cyanide into a cyanate, it is important to include the additional quantity implicated in the oxidation of the cation in the case of metallic cyanides such as copper (I) cyanide:

$$2\,CuCN + 3\,NaOCl + H_2O \longrightarrow Cu(OCN)_2 + Cu(OH)_2 + 3\,NaCl$$

$$Cu^I \xrightarrow{\ {}^-OCl\ } Cu^{II}$$

The particular case of cyanogen halides

Examples.

Cyanogen bromide	Br—CN
Cyanogen chloride	Cl—CN

Cyanogen halides, when dissolved in water, liberate hydrogen cyanide with a consequent risk of explosion.

The cyanogen halide must be dissolved with precaution in a cold solution of 10% sodium hydroxide to which is then added, while stirring, an excess of a commercial solution of sodium hypochlorite. After stirring overnight, the solution may be stored for waste disposal.

Remarks

The inhalation of a moderate quantity of hydrogen cyanide ($300\,mg/m^3$ or $270\,ppm$) is lethal within 5 minutes. Hydrogen cyanide is readily absorbed by mucous membranes (e.g., eyes) or lesioned skin (e.g., wounds). Similarly, ingestion of 50–100 mg of alkaline cyanide rapidly leads to anoxia and death.

Cyanogen halides are also very toxic (LD_{50} = 13 mg/kg for ClCN). At lower doses, these derivatives of hydrogen cyanide result in respiratory (dyspnoea), digestive (vomiting), and nervous disorders.

Cyanogen halides are sensitive to heat and must be kept in the refrigerator. Even under these conditions, the stability of cyanogen bromide is limited and spontaneous explosions of bottles have been reported.

In the case of frequent manipulations involving hydrogen cyanide or of media likely to liberate it, an atmospheric detection system linked to an alarm must be utilized and a medical doctor informed.

2.5 Destruction of Hydrofluoric Acid, Fluorides, and Their Derivatives

Examples

Hydrofluoric acid	HF
Sodium fluoride	NaF
Potassium fluoride	KF
Sodium fluoroborate	$NaBF_4$
Boron trifluoride	BF_3

Principle of destruction

Hydrofluoric acid is precipitated as calcium fluoride by addition of a solution of calcium hydroxide (slaked lime):

$$2\,HF + Ca(OH)_2 \longrightarrow CaF_2\downarrow + 2\,H_2O$$

Alkaline fluorides in aqueous solution are precipitated as calcium fluoride by the addition of an excess of a calcium chloride solution:

$$2\,F^- + CaCl_2 \longrightarrow CaF_2\downarrow + 2\,Cl^-$$

Boron trifluoride is first hydrolyzed in water and then treated with a solution of calcium chloride:

$$2\,BF_3 + 3\,CaCl_2 + 6\,H_2O \longrightarrow 3\,CaF_2\downarrow + 2\,B(OH)_3 + 6\,HCl$$

Destruction

Dissolve the fluorinated compound in excess water. Neutralize with sodium carbonate or first add a small quantity of hydrochloric acid followed by sodium carbonate until neutrality is attained. Add an excess of a calcium chloride solution in order to precipitate all the fluorides. Set aside. Filter. The collected precipitate is eliminated in the same manner as other solid wastes.

2.6 Destruction of Mineral Acids

Examples

Hydrochloric acid HCl
Hydrobromic acid HBr
Perchloric acid $HClO_4$

Nitric acid HNO_3
Hydroiodic acid HI
Sulfuric acid H_2SO_4

Principle of destruction

Progressive neutralization with a mineral base: alkaline bicarbonate ($NaHCO_3$), alkaline carbonate (Na_2CO_3), sodium hydroxide (NaOH), or potassium hydroxide (KOH).

Example: neutralization of a hydroacid HX (X=Cl, Br, I) by sodium carbonate

$$2\,HX + Na_2CO_3 \longrightarrow 2\,NaX + CO_2 + H_2O$$

Special precautions

Protective goggles
Rubber gloves
Safety shield (if necessary)

Elimination of spills

Cover the contaminated surface with a large excess of solid sodium bicarbonate or with a mixture of equal proportions of sodium carbonate and fused lime. Proceed with caution (exothermic reaction). Pour the mixture in small portions into a large excess of water (verify that the pH $\cong$ 7). Store for proper disposal.

Destruction of stocks

Into a large vessel containing an excess of cold 10% sodium hydroxide solution, carefully pour, while stirring, the acid to be destroyed. Monitor the temperature and, when all has been added, the pH. Store for proper disposal.

2.7 Destruction of Organic Acids

Examples

Carboxylic acids

Formic acid	Valeric acid
Acetic acid	Acrylic acid
Propionic acid	Methacrylic acid
n-Butyric acid	Pyruvic acid

Halogeno carboxylic acids
    Monochloroacetic acid   Dichloroacetic acid
    Trichloroacetic acid

Sulfonic acids
    Methanesulfonic acid   Benzenesulfonic acid
    p-Toluenesulfonic acid

Substituted phosphoric acids, etc.

Principle of destruction

Progressive neutralization with a mineral base: alkaline bicarbonate ($NaHCO_3$), alkaline carbonate (Na_2CO_3).

Example: neutralization of formic acid y sodium bicarbonate

$$H-\underset{\underset{O}{\|}}{C}-O-H + NaHCO_3 \longrightarrow H-\underset{\underset{O}{\|}}{C}-O^- Na^+ + CO_2 + H_2O$$

Special precautions

Identical to those proposed for mineral acids.

Elimination of spills

Cover the contaminated surface with a large excess of sodium bicarbonate or sodium carbonate. Proceed as in the case of mineral acids.

Destruction of stocks

Use the technique described for mineral acids.

2.8 Destruction of Oxalic Acid

Principle of destruction

Oxalic acid and the oxalates are irritants and nephrotoxins which can be degraded by heating in the presence of concentrated sulfuric acid:

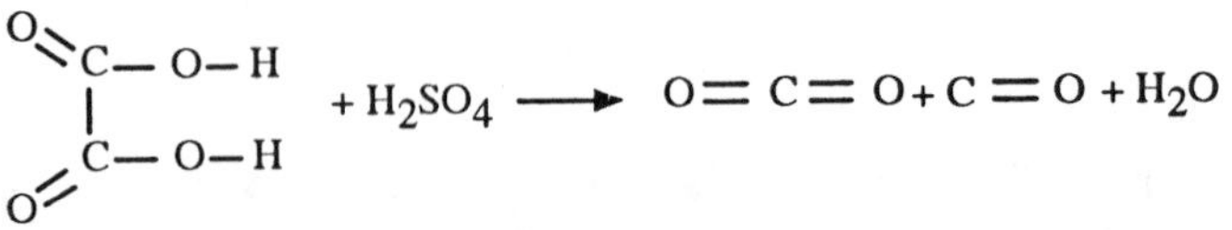

$$O=C-O-H \quad + H_2SO_4 \longrightarrow \quad O=C=O+C=O+H_2O$$
$$C-O-H$$

Method of destruction in the presence of sulfuric acid

In a round-bottom flask equipped with a stirrer, a thermometer, and a gas outlet, place:

  –25 ml of concentrated H_2SO_4
  –5 g of oxalic acid or oxalate

Heat carefully at 80–100°C for 30 minutes. After cooling, the reaction mixture is carefully poured into a large excess of ice-cold water. The solution is neutralized by the addition of sodium carbonate and then properly disposed.

Other methods of destruction

Oxalic acid and oxalates may be destroyed by oxidation with an acid solution of potassium permanganate held at 80°C for 1 hour.

  Oxalic acid can also be precipitated in water solution by the addition in excess of an aqueous solution of calcium chloride. The insoluble calcium oxalate is filtered and eliminated in the same manner as other solid waste.

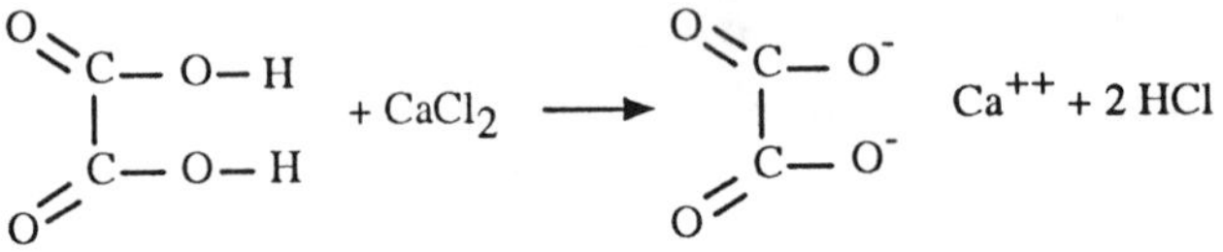

$$O=C-O-H \quad + CaCl_2 \longrightarrow \quad O=C-O^- \quad Ca^{++} + 2\,HCl$$
$$C-O-H \qquad\qquad\qquad C-O^-$$

2.9 Destruction of Simple Aldehydes and Ketones

Examples

Aldehydes
  Formaldehyde (formalin)
  Acetaldehyde
  Propionaldehyde
  n-Butyraldehyde
  Benzaldehyde
  Furfuraldehyde (furfural)
  Acrolein
  Crotonaldehyde
  Glutaraldehyde

Ketones
  Acetone
  Methyl ethyl ketone
  Methyl n-butyl ketone
  Methyl vinyl ketone
  Cyclopentanone
  Cyclohexanone

Many aldehydes (e.g., formaldehyde, acrolein, crotonaldehyde) are irritants, sensitizers and, a few, are carcinogenic.*

Ketones are relatively nontoxic except for methyl n-butyl ketone and 2,5-hexanedione, which are peripheral neurotoxins.

Principle of destruction

Provided an appropriate facility is available, the best technique for the elimination of aldehydes and ketones is incineration. Ketones used as solvents (e.g., acetone, methyl ethyl ketone, cyclohexanone) may be eliminated with other nonhalogenated solvents. The majority of aldehydes and ketones are oxidized to carboxylic acids by treatment with potassium permanganate:

$$3\,R-\underset{\underset{O}{\|}}{C}-H \;+2\,KMnO_4 \longrightarrow 3\,R-\underset{\underset{O}{\|}}{C}-OH + 2\,MnO_2 + H_2O$$

α,β-Unsaturated aldehydes (e.g., acrolein) and α,β-unsaturated ketones (e.g., methyl vinyl ketone) are also cleaved by $KMnO_4$:

$$3\,R-\underset{\underset{O}{\|}}{C}-CH\!=\!CH\!-\!R + 8\,KMnO_4 \longrightarrow 3\,R-\underset{\underset{O}{\|}}{C}-C\overset{OK}{\underset{O}{\diagup\!\!\diagdown}} + 3\,R-\underset{\underset{O}{\|}}{C}-OK$$

$$+\,2KOH + 8\,MnO_2 + 2\,H_2O$$

In this case, an excess of $KMnO_4$ must be used to ensure complete destruction of these compounds.

In the case of methyl ketones $R\text{---}CO\text{---}CH_3$ such as methyl n-butyl ketone, destruction is achieved by contact with an excess of a sodium hypochlorite solution.

$$R-\underset{\underset{O}{\|}}{C}-CH_3 + 3\,NaOCl \longrightarrow R-\underset{\underset{O}{\|}}{C}-ONa + CHCl_3 + 2\,NaOH$$

Aldehydes and certain unhindered ketones (e.g., acetone, methyl ethyl ketone, cyclopentanone, cyclohexanone) form a bisulfite addition product upon treatment with an excess of sodium bisulfite ($NaHSO_3$). Generally poorly soluble in concentrated media, the bisulfite addition product is soluble in excess water:

$$\underset{R'}{\overset{R}{\diagdown}}C\!=\!O \;\;+NaHSO_3 \longrightarrow \underset{R'}{\overset{R}{\diagdown}}\underset{\diagdown}{\overset{\diagup SO_3Na}{C}}_{OH}$$

Commercial sodium bisulfite exists both in an anhydrous form (sodium

---

*Formaldehyde, acetaldehyde, acrolein, and crotonaldehyde have been shown to be carcinogenic in animals.

metabisulfite: $Na_2S_2O)_5$ and in aqueous solution (with a minimum SO_2 concentration of 58.5%).

Special precautions

Safety goggles
Rubber gloves

Oxidative destruction using an acidic solution of potassium permanganate

Add to a three-necked, round-bottom flask equipped with a stirrer, a condenser, a dropping funnel and a thermometer:

 −100 ml of water
 −0.1 mM of aldehyde

While stirring, approximately 30 ml of a solution of 12.6 g of potassium permanganate (0.08 mol, corresponding to 20% excess) in 250 ml of water is added dropwise over 15 minutes. The temperature is monitored.

If, after the addition has been terminated, the purple solution turns completely colorless, more potassium permanganate solution must be added until a stable purple color is obtained. Heat this solution at 70–80°C for 1 hour while stirring. The solution, cooled to room temperature, is acidified with 6 N H_2SO_4. In order to reduce the excess $KMnO_4$ to MnO_2, 8.3 g (0.08 mol) of solid sodium metabisulfite (or the equivalent of a solution of sodium bisulfite) is added over several minutes, the temperature being kept at 20–40°C. If the resulting solution is not colorless, more sodium bisulfite must be added. Store for disposal.

With unsaturated aldehydes (such as acrolein) or unsaturated ketones, it is necessary to use 4 mol of $KMnO_4$ (+20% excess) per mole of compound.

2.10 Destruction of Diverse Oxygenated Compounds

Principle of destruction

Low molecular weight alcohols commonly used in laboratories as solvents (e.g., methanol, ethanol, isopropanol, etc.) are easily biodegraded in aqueous media. Water-soluble alcohols, as well as higher molecular weight alcohols, should be eliminated with nonhalogenated liquid wastes. Ethers and esters may be handled in the same manner.

Generally toxic, phenols should be destroyed by oxidation, usually with aqueous hydrogen peroxide in the presence of a metal catalyst. The use of sodium hypochlorite is to be avoided owing to the formation of chlorophenols that pollute the aquatic environment.

Destruction of phenols by oxidation

To a three-necked, round-bottom flask equipped with a magnetic stirrer, a thermometer, a dropping funnel, and a condenser with a gas outlet, add 750 ml of water and 47 g of phenol (0.5 mol). Stir until completely dissolved and then add in small portions 25.5 g of ferrous sulfate heptahydrate (0.085 mol). Adjust the pH of the solution to 5–6 by the addition of dilute sulfuric acid. While stirring, slowly add 410 ml of 30% aqueous hydrogen peroxide (4 mol) drop-wise over 1 hour. Because the oxidation of phenol by hydrogen peroxide is exothermic, the temperature of the mixture must be maintained at 50–60°C by controlling the rate of addition of the peroxide.

After complete addition of the hydrogen peroxide, the mixture is stirred for 2 hours while allowing it to cool to room temperature. After standing over-night, the mixture can then be stored for disposal.

This procedure may be applied to cresols and naphthols.

2.11 Destruction of Alkali Amides

Examples

Lithium amide	$LiNH_2$
Potassium amide	KNH_2
Sodium amide	$NaNH_2$

Principle of destruction

Alkaline amides may be destroyed by the action of excess solid ammonium chloride, ammonia being liberated in the process:

$$NaNH_2 + NH_4Cl \longrightarrow NaCl + 2 NH_3$$

Potassium amide, the most reactive, as well as sodium amide, may also be destroyed by careful addition to cold alcohols (ethanol, isopropanol):

$$KNH_2 \rightarrow C_2H_5OH \rightarrow C_2HOK + NH_3$$

Remarks

Commercial sodium amide is an unstable product which, during storage, absorbs moisture and oxygen and thus decomposes, becoming covered with a yellowish film (peroxide derivatives?). This "aged" product becomes very unstable and may spontaneously explode. When this "aged" amide is immersed in water, no reaction is observed at first. However, the reaction quickly becomes violent and explosive. *This amide covered with a yellowish film must be manipulated with extreme caution.* It can be destroyed by controlled inciner-ation.

Special precautions

Safety goggles
Rubber gloves
Polycarbonate safety shield
Manipulation under a well-ventilated fume-hood

Destruction by an alcohol

Cover the amide with a dry hydrocarbon (e.g., white spirit, toluene, etc.). Carefully add 95% ethanol while stirring. Dispose of with other non-halogenated products.

Destruction using excess ammonium chloride

To an excess of solid ammonium chloride, add the alkaline amide in small portions. Let stand overnight and dispose of with other solid wastes.

Destruction within a reaction mixture

At the end of the reaction period (ethynylation, etc.), add an excess of solid ammonium chloride to the mixture. Allow to stir and then evaporate the solvent (liquid ammonia). Carefully add ice water to the residue. Extract the reaction product with a solvent. Store for proper disposal.

2.12 Destruction of Azides and Diazo Compounds

Examples

Hydrazoic acid	HN_3
Lithium azide	LiN_3
Sodium azide	NaN_3
Lead azide	$Pb(N_3)_2$
Cupric azide	$Cu(N_3)_2$
Diazomethane	$^-N{=}^+N{=}CH_2$

Principle of destruction

Metallic azides are quantitatively oxidized in the presence of ceric salts (ceric ammonium nitrate), liberating nitrogen.

$$2\ ^-N = N^+{=\!=} N^- + 2\,Ce^{IV} \longrightarrow 2\,(^-N = N^+{=\!=} N^\bullet) + 2\,Ce^{III}$$

$$\downarrow$$

$$N \equiv N$$

Azides may also be destroyed by reaction with nitrous acid:

$$2\,NaN_3 + 2\,HNO_2 \longrightarrow 3\,N_2 + 2\,NO + 2\,NaOH$$

Special precautions

Safety goggles
Rubber gloves
Polycarbonate safety shield
Work with small quantities, under a very efficient fume-hood

Remark

Hydrazoic acid and alkaline azides are highly toxic compounds.

Sodium azide (NaN_3) is not explosive unless heated close to its decomposition temperature (300°CC).

Heavy metal azides (Pb, Ag, Hg, Cu) and alkyl azides (R—N_3) are explosive. Exposing these to heat, friction, or mechanical shock must be avoided.

Diazomethane and diazo compounds (R—N_2) are also very toxic substances (carcinogenic in animals) and explosive in the dry form.

Elimination of spills

Cover the contaminated area with a 10% aqueous solution of ceric ammonium nitrate $(NH_4)_2Ce(NO_3)_6$. Allow to react, then dilute with water and store for disposal.

Destruction of alkaline azide stocks

Ceric ammonium nitrate method.   To a cold solution of the alkaline azide or of hydrazoic acid, slowly add, while stirring, an excess of 10% aqueous ceric ammonium nitrate solution. Dilute with excess water and store for disposal.

Nitrous acid method.   Work under a well-ventilated fume-hood. To a cold solution containing less than 5% azide, add, while stirring, an excess of a 20% aqueous sodium nitrite solution (1.5 g of sodium nitrite per gram of sodium azide). Very slowly add a cold solution of 20% sulfuric acid such that the azide solution becomes weakly acidic (pH 5–6). Beware of the formation of toxic nitrogen oxide vapors. The destruction of the azide is complete when the solution turns iodinated starch paper blue (indicating the presence of excess nitrite). Store for disposal.

Remark

If sulfuric acid is added to the azide solution before the addition of the nitrite, highly toxic hydrazoic acid may be liberated.

Destruction of lead azide

Dissolve the lead azide in 10% aqueous ammonium acetate. Cautiously add a 10% solution of potassium dichromate until complete precipitation of lead chromate (carcinogenic) is achieved. Filter, wash, and eliminate with other solid wastes.

Destruction of hydrazoic acid solutions

Hydrazoic acid (HN_3) is a highly unstable and highly toxic compound (its toxicity is identical to that of hydrogen cyanide) which should never be isolated. Solutions of hydrazoic acid in benzene or chloroform are relatively stable.

A benzene or chloroform solution of hydrazoic acid placed in a separatory funnel is decomposed by adding sufficient water such that the aqueous phase contains less than 5% HN_3. The aqueous phase is neutralized with aqueous sodium hydroxide, then separated from the organic phase and treated with ceric ammonium nitrate or with sodium nitrite in an acidic medium.

Always work under a well-ventilated fume-hood.

Destruction of diazomethane

Diazomethane may be destroyed by the slow, cautious addition of acetic acid. The methyl acetate derivative formed may be disposed of with other solvents.

$$^-N=N^+=CH_2 \ + \ CH_3-\overset{\overset{\displaystyle O}{\|}}{C}-O-H \ \longrightarrow \ CH_3-\overset{\overset{\displaystyle O}{\|}}{C}-O-CH_3 \ + \ N\equiv N$$

2.13 Destruction of Mineral Bases

Examples

Ammonium hydroxide	NH_4OH
Calcium hydroxide (slaked lime)	$Ca(OH)_2$
Lithium hydroxide	$LiOH$
Sodium hydroxide	$NaOH$
Potassium hydroxide	KOH

Principle of destruction

Progressive neutralization with dilute mineral acid (HCl or H_2SO_4) leading to formation of a water-soluble salt.

Example: neutralization of sodium hydroxide by hydrochloric acid

$$NaOH + HCl \longrightarrow NaCl + H_2O$$

Special precautions

Protective goggles
Rubber gloves
Safety shield (if necessary)

Elimination of spills

Cover the contaminated area with a slight excess of a cold solution of 6 M hydrochloric acid or of sulfuric acid diluted by half. Proceed with caution (exothermic reaction). Monitor the pH, which should eventually be close to neutral. Collect the mixture.

Destruction of stocks

Into a large vessel containing an excess of cold 6 M HCl solution, carefully pour in, while stirring, the base to be destroyed. Monitor the temperature and, after completion of the addition, the pH. Store for waste disposal.

2.14 Recovery of Noble Metal Catalysts

Noble metal catalysts (containing platinum, palladium, rhodium, ruthenium, etc.) may be recovered and stored in an aqueous medium in tightly stoppered glass flasks. The regeneration of these catalysts requires the specific techniques described in *Handbook of preparative inorganic chemistry*, by H. L. Grube. 2nd Ed. Vol. 2, G. Braver, Ed., Academic Press, New York (1965), pp. 1560–1605.

Remarks

After hydrogenation, catalysts remain saturated with hydrogen and are extremely reactive. They may spontaneously ignite if exposed to air and they must never be allowed to become dry, especially on a combustible surface (e.g., paper filters, etc.).

At the end of a hydrogenation, the reaction vessel must be carefully purged with inert nitrogen. After the catalyst has been filtered and washed with the solvent used for the hydrogenation, it is carefully washed with water, collected and stored under moist conditions.

A noble metal catalyst should never be directly introduced in a reaction mixture already under a hydrogen atmosphere.

Destruction of Raney nickel-type catalysts

After hydrogenation, Raney nickel-based catalysts are extremely reactive and may easily ignite in air. After copious washing with water, they can be collected and suspended in excess water contained in a 3 l flask equipped with an efficient stirrer. For approximately 10 g of catalyst, 200 ml of water are required. Using a dropping funnel, gradually add 800 ml of 1 N HCl until the nickel has completely dissolved. Monitor the temperature.

The nickel may also be precipitated as a sulfide using the method described on p. 000.

Remark

Zinc-based catalysts (used for Clemmensen reductions, etc.) which are also very pyrophoric, may be recovered using this technique.

2.15 Destruction of Basic Organic Compounds

Examples

Aliphatic amines (mono, di, tri)
   Methylamine
   Ethylamine
   n-Propylamine
   Isopropylamine
   Butylamines
   Cyclohexylamine

Alicyclic amines
   Pyrrolidine
   Piperidine
   Morpholine

Aromatic amines
   Aniline
   Toluidines (o, m, p)
   Xylidines (o, m, p)

Polyamines
   Ethylenediamine

Pyridine bases
   Pyridine
   Collidines
   Picolines
   Quinoline
   Isoquinoline

Substituted hydrazines
   Methylhydrazine
   Dimethylhydrazine
   Phenylhydrazine

Alkaloids
   Nicotine
   Piperine

Amino alcohols
   Ethanolamines (mono, di, tri)

Principle of destruction

The basic compound may be transformed into a water-soluble salt by the action of sulfuric acid (H_2SO_4), sodium bisulfate ($NaHSO_4$), or sulfamic acid

(H_2NSO_3H).

Sulfamic acid

Example: elimination of a primary amine by formation of a water-soluble sulfate in the presence of sulfuric acid

$$R\text{-}NH_2 + H_2SO_4 \longrightarrow RNH_3^+ HSO_4^-$$

This technique cannot be applied to very weakly basic compounds (e.g., triphenylamine). Moreover, certain amines, such as diphenylamine (C_6H_5—NH—C_6H_5) form insoluble sulfates.

Special precautions

Protective goggles
Butyl rubber gloves

It should be noted that aromatic amines penetrate skin with ease. These procedures must not be used on carcinogenic compounds of this class (e.g., 2-naphthylamine, benzidine, etc.) which should be destroyed by specific techniques.

Elimination of spills

Sodium bisulfate technique.   Cover the contaminated area with a large excess of solid sodium bisulfate. Proceed with caution when dealing with strongly basic compounds (exothermic reaction). Spray water onto the mixture.

Sulfuric acid technique.   Carefully neutralize the basic compound with a solution of sulfuric acid diluted by half. Check the final pH, which should be close to neutral.

Sulfamic acid technique.   Cover the contaminated area with a large excess of solid sulfamic acid. Pour the mixture in small portions into a large excess of water. Check the final pH ($\cong 7$).

Destruction of stocks

Sodium bisulfate technique.   Into a large vessel containing an excess of a cold solution of sodium bisulfate (soluble in two parts of water at 20°C), add the base to be destroyed. Monitor the temperature and, after completion of the addition, the pH. Dilute with excess water. Store for disposal.

Sulfuric acid technique.   The same technique as with sodium bisulfate, using H_2SO_4 diluted by half.

Sulfamic acid technique.   Into a large vessel containing excess saturated sulfamic acid solution (soluble in 6.5 parts of water at 20°C), add, while stirring and with cooling, the basic compound. Monitor the temperature and, after completion of the addition, the pH (which should be nearly neutral). Dilute with an excess of water. Store for disposal.

Remark

This technique, very appropriate for alkaloids (biologically active compounds) and for heterocyclic bases (e.g., pyridine) may also be used for certain amides (e.g., formamide).

Destruction of aromatic amines

Numerous aromatic amines (e.g., aniline, N,N'-dimethylaniline, o-toluidine, chloroanilines, p-aminophenol, p-nitroaniline, p-phenylenediamine, diphenyl-amine, 1-naphthylamine, etc.) are toxic to red blood cells, resulting in severe methemoglobin formation. Many aromatic amines (e.g., 4-aminobiphenyl, 2-naphthylamine, benzidine, etc.) are also powerful bladder carcinogens in humans. These carcinogenic aromatic amines should preferably be destroyed by permanganate oxidation in sulfuric acid.

In general, aromatic amines may be eliminated after transformation into their corresponding aromatic hydrocarbons. This reduction is effected by first converting the aromatic amine into a diazonium salt by the action of sodium nitrite in the presence of hydrochloric acid. The diazonium salt is then treated with excess hypophosphorous acid.

$$ArNH_2 + NaNO_2 + 2\,HCl \longrightarrow (\,Ar - N^+ \equiv N\,Cl^-\,) \; + 2\,H_2O + NaCl$$
$$\text{diazonium chloride}$$

$$(\,Ar - N^+ \equiv N\,Cl^-\,) \; + H_3PO_2 + H_2O \longrightarrow ArH + H_3PO_3 + HCl + N \equiv N$$

Method.   To a 1 l, three-necked round-bottom flask equipped with an efficient stirrer, a dropping funnel, a thermometer, and a gas outlet, add:

- −25 ml of water
- −75 ml of 36% HCl
- −0.2 mol of aromatic amine (added slowly)

The mixture is stirred until complete dissolution of the amine hydrochloride is achieved. The temperature is then brought to between −5 and −10°C by use of a cooling bath. A solution of 15 g (0.211 mol) of sodium nitrite in 35 ml of water is added dropwise such that the temperature of the reaction mixture never exceeds 0°C. After completion of the addition, the mixture is allowed to stir for another 30 minutes. To this solution of arenediazonium chloride, maintained at −5°C, is added 416 ml (4 mol) of 50% hypophosphorous acid (maintained at 0°C) over 15 minutes. Stirring is continued for 1 hour after which the mixture, allowed to come to room temperature, is left to stand for 24 hours. The mixture is then poured into a separatory funnel and extracted twice with 100 ml of toluene. The combined organic extracts are eliminated with other waste solvents. Store for disposal.

2.16 Destruction of Hydrogen Peroxide and Derivatives

Examples

Hydrogen peroxide	Hydroperoxides
	t-Butyl hydroperoxide
Peroxides	Peresters
Sodium peroxide	t-Butyl perbenzoate
Alkyl peroxides	Alkyl peroxydicarbonate
Peracids	Peranhydrides
Peracetic acid	Trifluoroperacetic anhydride
Perbenzoic acid	Benzoyl peroxide
Perphthalic acid	

Principle of destruction

Hydrogen peroxide and its principal derivatives (e.g., hydroperoxides, peroxides, peracids, persalts, peresters, peranhydrides, etc.) may be destroyed by various powerful reducing agents (e.g., sodium sulfite, sodium bisulfite, ferrous sulfate, etc.).

$$\text{R-O-O-H} + 2\,Fe^{2+} + 2\,H^+ \longrightarrow \text{R-OH} + 2\,Fe^{3+} + H_2O$$

Organic derivatives of hydrogen peroxide (e.g., hydroperoxides, peroxides, etc.) are easily destroyed by incineration. In the case of peresters and peranhydrides, hydrolysis using strong, hot mineral bases (e.g., sodium hydroxide, potassium hydroxide) is generally effective.

In general, derivatives of hydrogen peroxide such as primary and secondary hydroperoxides and peroxides are destroyed in alkaline media. For example, certain hydroperoxides in strong alkali are transformed into alcohols, and sometimes into carbonyl derivatives, when heated.

$$2\,R - O - OH \xrightarrow{\ OH^-\ } 2\,R - O - H + O_2$$

Peranhydrides that are poorly soluble in water, such as benzoyl peroxide, react very slowly. In this case, destruction is achieved by treatment with a solution of an alkali iodide in hydrochloric acid.

$$\underset{\text{R}-\overset{\displaystyle \|}{\underset{}{C}}-O-O-\overset{\displaystyle \|}{\underset{}{C}}-\text{R}}{\overset{O\qquad\qquad O}{}} + 2\,NaI \longrightarrow 2\,R-\overset{\displaystyle \overset{O}{\|}}{C}-ONa + I_2$$

Special precautions

Protective goggles
Rubber gloves
Safety shield

Elimination of spills

Spilled liquids may be soaked up with large quantities of vermiculite. Carefully place the mixture in a plastic container with the help of a plastic spoon or scoop. Burn the material in an incinerator, as provided by state/federal agencies.

Destruction of stocks

Hydrogen peroxide, water-soluble peracids.   Dilute with 20 parts of water. Neutralize with 10% sodium hydroxide.

Mineral peroxides (e.g., sodium peroxide, etc.).   Pour in small portions into a large excess of aqueous sodium sulfite solution (50% excess).

Hydroperoxides.   Carefully pour into a large excess of 10% sodium hydroxide and slowly add an excess of a sodium sulfite solution.

Peroxides.   Except by incineration, the destruction of peroxides is very difficult and requires the use of powerful reducing agents (e.g., stannous chloride, zinc in acid medium, $LiAlH_4$, or alkaline iodides in acid medium) often accompanied by heating.

In a round-bottom flask equipped with a stirrer and a condenser, place:

−70 ml of acetic acid
−0.022 mol (10% excess) of potassium iodide

Stir until completely dissolved, then add:

−1 ml of 36% HCl

Then add to this solution:

−0.01 mol of alkyl peroxide (R—O—O—R)

Slowly heat over 30 minutes to 90–100°C and leave at this temperature for 5 hours. Check for the absence of peroxides by using the acidic potassium iodide test (3 g of potassium iodide in 50 ml of acetic acid containing 2 ml of 36% HCl). When peroxides are no longer present, cool the solution, dilute with water, and store for disposal.

Peresters, peranhydrides.   Carefully pour the perester (e.g., t-butyl perbenzoate, etc.) or the peranhydride (e.g., peracetic anhydride, etc.) into an excess of a cold solution of 20% methanolic sodium hydroxide. Store for disposal.

As an example, dissolve 10 g of di-t-butyl peroxyoxalate in 1 l of methanol.

Cool the solution to 0°C and carefully add 50 ml of a 30% solution of sodium hydroxide. Store for disposal.

2.17 Destruction of Alkyl, Allyl, or Benzyl Halides

Examples

$$\text{Alkyl, allyl, benzyl} \quad\begin{cases} \text{chlorides} \\ \text{bromides} \\ \text{iodides} \end{cases}$$

Benzyl fluoride

The methods of destruction described may also be applied to other alkylating agents:

Dialkyl sulfates (methyl, ethyl, etc.)
Alkyl methanesulfonates (methyl, ethyl, etc.)
Trialkyl phosphates (methyl, ethyl, etc.)
Triaryl phosphates (phenyl, cresyl, etc.)

Principle of destruction

Organic halides may be destroyed by incineration in an area reserved for this purpose. Before incineration, the halogenated derivative should be well mixed with calcium carbonate or with slaked lime in order to trap the acids formed during combustion.

In the laboratory, sufficiently reactive halides (e.g., alkyl, allyl, benzyl halides) may be decomposed by hydrolysis in a strongly alkaline medium, with formation of the corresponding alcohols.

$$\text{R-X} + \text{KOH} \longrightarrow \text{R-OH} + \text{KX}$$

Depending on the structure of the halide, this simple nucleophilic substitution may be accompanied by secondary reactions: formation of ethers, alkenes, and so on.

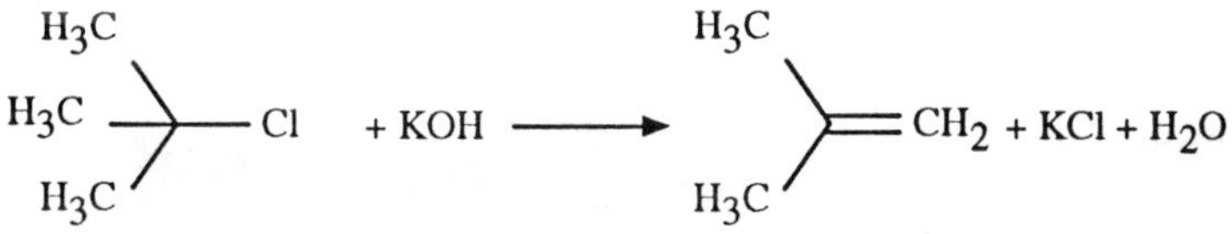

t-butyl chloride   Isobutene

Hydrolysis in strongly basic media may be used for most primary, secondary, or tertiary alkyl chlorides, bromides, and iodides. This method is also applicable to allyl and benzyl halides (including fluorides). However, this technique is not sufficient to hydrolyze vinyl or aryl halides or most fluorinated derivatives.

Remark

This technique may, however, be used to eliminate numerous classes of alkylating agents: sulfates, sulfonates, phosphates, α or β halogenated ethers, epoxides, aziridines, and so on. More nucleophilic compounds (e.g., sulfates, alkylsulfonates, haloethers, etc.) may be more easily hydrolyzed with 10% aqueous sodium hydroxide.

Special precautions

Protective goggles
Rubber gloves

Destruction

Incineration.   Mix the halide with vermiculite in the presence of excess sodium carbonate and calcium hydroxide (slaked lime). Burn the mixture in an incinerator generally reserved for halogenated derivatives, as provided by state/federal agencies.

Alkaline hydrolysis.   In a 1 l, round-bottom flask equipped with a magnetic stirrer, a thermometer, a condenser, and a dropping funnel, place:

   −79 g of 85% potassium hydroxide pellets (1.2 mol)
   −315 ml of 95% ethanol

The temperature will rise to 55°C as the mixture dissolves. Heat to reflux. Add the halide dropwise with stirring (either directly if the halide is a liquid or as a solution in 95% ethanol if it is a solid). The rate of addition should be adjusted such that reflux is just maintained. After rinsing the dropping funnel with a small quantity of 95% ethanol, reflux is continued while stirring for 2 hours. The potassium halide gradually precipitates. Allow the mixture to cool to room temperature. Store for disposal.

2.18 Destruction of Acid Halides and Anhydrides

Examples

Acetyl bromide	Crotonyl chloride
Acetyl chloride	Oxalyl chloride
Trichloroacetyl chloride	Methanesulfonyl chloride
Trifluoroacetyl chloride	Benzenesulfonyl chloride
Propionyl chloride	p-Toluenesulfonyl chloride
Benzoyl chloride	
Acetic anhydride	
Propionic anhydride	

Principle of destruction

Acid halides and anhydrides are hydrolyzed by mineral bases: sodium carbonate (Na_2CO_3), sodium bicarbonate ($NaHCO_3$), ammonium hydroxide (NH_4OH), and sometimes sodium hydroxide (NaOH) or potassium hydroxide (KOH) for certain derivatives such as benzoyl chloride. The use of dilute solutions of ammonium hydroxide allows mild and complete destruction of most acid halides and anhydrides. The reaction of water with low molecular weight acid halides and anhydrides is very exothermic and must be carefully controlled.

$$\underset{R\,-\,C\,-\,X\,+\,2\,NaOH}{\overset{\overset{\textstyle O}{\|}}{}} \longrightarrow \underset{R\,-\,C\,-\,ONa\,+\,NaX\,+\,H_2O}{\overset{\overset{\textstyle O}{\|}}{}}$$

Potassium salts being more water-soluble than sodium salts, hydrolysis with potassium hydroxide is generally preferred.

Special precautions

Protective gastight goggles
Rubber gloves
Safety shield (if necessary)

Acid halides and anhydrides are generally irritants and often lachrimators.

Elimination of spills

Cover the contaminated area with a large excess of powdered sodium bicarbonate. Proceed with caution (exothermic reaction). Pour the mixture into a large excess of water.

Destruction of stocks of easily hydrolyzed halides

Into a large vessel containing an excess of an ice-cold alkaline solution (e.g., Na_2CO_3, $NaHCO_3$, NH_4OH, etc.), carefully pour the acid halide or anhydride to be destroyed in small portions while stirring. Monitor the temperature of the reaction mixture. With carbonate or bicarbonate, foaming and liberation of large quantities of gas (CO_2) is observed. After complete addition, check the pH (which should be alkaline). Allow the mixture to stir for 1 hour. Store for disposal.

Destruction of stocks of halides difficult to hydrolyze (e.g., benzoyl chloride, p-toluenesulfonyl chloride, etc.)

In a three-necked flask equipped with a stirrer, a condenser, a dropping funnel, and a thermometer, place:

 −600 ml of 2.5 N sodium hydroxide (i.e., 50% excess of base)

Add the halide very slowly while stirring such that the reaction remains exothermic. Heat to 90°C in order to dissolve all the halide. Once the solution is clear, allow it to come to room temperature. Neutralize to pH 7 with dilute HCl. Dilute with water and store for disposal.

2.19 Destruction of Mineral Halides

Examples

Anhydrous aluminum chloride	Germanium tetrachloride
Aluminum bromide	Silicon tetrachloride
Antimony trichloride	Tin tetrachloride (stannic chloride)
Antimony pentachloride	Titanium tetrachloride
Ferric chloride (iron perchloride)	Zirconium tetrachloride

Principle of destruction

Reactive mineral halides are hydrolyzed in the presence of sodium bicarbonate or ammonium hydroxide. The use of an excess of base is necessary in order to dissolve the metallic hydroxide formed.

Example.

$$2\,AlCl_3 + 3\,H_2O \xrightarrow{\ OH^-\ } Al_2O_3 + 6\,HCl$$

Special precautions

Protective goggles
Rubber gloves
Safety shield (if necessary)

Many reactive mineral halides are powerful irritants and may release very corrosive acid vapors.

Elimination of spills

Cover the contaminated area with a large excess of powdered sodium bicarbonate. Proceed with caution (exothermic reaction). Pour the mixture into a large excess of 50% sodium hydroxide solution. If necessary, add more sodium hydroxide to dissolve all the precipitate that may form.

Destruction of stocks

Into a large vessel containing a large excess of an ice-cold solution of 6 M ammonium hydroxide, carefully pour the halide to be destroyed in small portions while stirring. Monitor the temperature. Ammonium halide fumes will be released. Stir for 1 hour. Verify the pH. Store for disposal.

Remark

Boron trihalides (BBr_3, BCl_3, BF_3) may also be destroyed by contact with alkali ($NaHCO_3$ or 2.5 N NaOH). Boron trifluoride (BF_3) may also be eliminated under the form of calcium fluoride.

2.20 Destruction of Organometallic Hydrides and Derivatives

Examples

Hydrides
    Lithium hydride LiH
    Potassium hydride KH
    Sodium hydride NaH
    Calcium hydride CaH_2

Borohydrides
    Lithium borohydride $LiBH_4$
    Potassium borohydride KBH_4
    Sodium borohydride $NaBH_4$

Aluminum hydrides
    Lithium aluminum hydride $LiAlH_4$

Arsines
    Arsine AsH_3

Boranes
    Diborane B_2H_6
    Pentaborane (9) B_5H_9
    Decaborane (14) $B_{10}H_{14}$

Germanes
    Germane GeH_4

Phosphines
    Phosphine PH_3
    Diphosphane $H_2P\!-\!PH_2$

Silanes
    Silane SiH_4

Organometallic compounds
    Magnesium: Grignard reagents $R\!-\!Mg\!-\!X$
    Aluminum: Trimethylaluminum $(CH_3)_3Al$
    Bismuth: Trimethylbismuth $(CH_3)_3Bi$
    Cadmium: Dimethylcadmium $(CH_3)_2Cd$
    Zinc: Dimethylzinc $(CH_3)_2Zn$

Principle of destruction

Most of these compounds often react violently with water (explosion, fire).

$$NaH + H_2O \longrightarrow NaOH + H_2$$

Water-sensitive compounds may be destroyed by contact with a higher alcohol such as *n*-butanol.

Example: destruction of diborane by *n*-butanol

Diborane reacts rapidly with *n*-butanol, forming *n*-butyl borate and liberating hydrogen. In the presence of water, *n*-butyl borate is hydrolyzed to boric acid and *n*-butanol.

$$B_2H_6 + 6\, C_4H_9OH \longrightarrow 2\,(C_4H_9O)_3B + 6\, H_2$$
$$\downarrow H_2O$$
$$6\, C_4H_9OH + 2\, H_2BO_3$$

Silane decomposes very slowly in water but more rapidly in alkaline media. Arsine (AsH_3) and phosphine (PH_3) are stable in water. Nonmetallic hydrides (SiH_4, AsH_3, PH_3) unreactive with water but very sensitive to oxidizing agents may be destroyed by oxidation (e.g., $KMnO_4$, copper sulfate, etc.).

Special precautions

Protective goggles
Rubber gloves
Safety shield

Many hydrides and organometallic compounds react violently with oxygen and may spontaneously ignite when in contact with air. In these cases, it is necessary to work under an inert atmosphere (nitrogen, argon).

Destruction

To a large vessel equipped with a stirrer and containing cold *n*-butanol, carefully add the substance to be destroyed in small portions. Caution! hydrogen is liberated; monitor the temperature of the reaction mixture. When reaction has stopped, store for disposal.

Borohydrides ($NaBH_4$, KBH_4, $LiBH_4$) may be destroyed by slow addition to an excess of ice-cold water. The use of dilute hydrochloric acid solution facilitates the destruction of poorly reactive compounds.

Destruction of lithium aluminum hydride

Lithium aluminum hydride ($LiAlH_4$) reacts violently with water and with many compounds that have a mobile hydrogen atom (e.g., alcohols, acids, etc.) or that are easily reduced (e.g., esters, etc.).

$$LiAlH_4 + 4 H_2O \longrightarrow LiOH + Al(OH)_3 + 4 H_2$$

Contact of solid lithium aluminum hydride with small quantities of water (humidity, etc.) may lead to spontaneous combustion. Simple handling may also provoke explosions; working under an inert atmosphere (nitrogen, argon) is thus strongly recommended. An excess of lithium aluminum hydride (usually two to four times the theoretical quantity) is generally used for the reduction of organic compounds and this excess must be destroyed at the end of the reaction. In large-scale operations, destruction of the hydride by addition of small quantities of ice-cold water is always very dangerous due to the liberation of large quantities of hydrogen that may burn or explode. In these cases it is preferable to use 10% solutions of sodium hydroxide or ammonium chloride. The alumina which precipitates is then easily removed by filtration. If the reaction products are sufficiently stable, excess $LiAlH_4$ may be destroyed by the addition of an ice-cold solution of 20% sodium hydroxide (5 N).

The best method of destroying excess $LiAlH_4$ appears to be that employing ethyl acetate. This ester destroys $LiAlH_4$ without liberating hydrogen; care must nevertheless be exercised since the reduction is exothermic.

$$4 \ CH_3 - \overset{\displaystyle O}{\overset{\displaystyle \|}{C}} - O - C_2H_5 + LiAlH_4 \longrightarrow C_2H_5 - O - Li + (C_2H_5O)_3Al + 4 \ C_2H_5OH$$

This technique of destruction must be avoided in the case of basic compounds (amines, alkaloids, etc.) due to possible secondary acetylation reactions with ethyl acetate. The method using 20% sodium hydroxide is then preferred.

Destruction of nonmetallic hydrides

Nonmetallic hydrides such as diborane (B_2H_6), hydrogen phosphide or phosphine (PH_3), hydrogen arsenide or arsine (AsH_3), and especially silane (SiH_4) are very sensitive to oxidizing agents and may spontaneously ignite in air (SiH_4, PH_3,* etc.). The self-ignition temperature of silane is inferior to ambient temperature and this gas may thus spontaneously burn in air and may, under certain conditions (presence of impurities), form explosive mixtures.

$$SiH_4 + 2 O_2 \longrightarrow SiO_2 + 2 H_2O + 246 \ Kcal.$$

Nonmetallic hydrides (B_2H_6, PH_3, AsH_3) may be destroyed by oxidation in the presence of an aqueous solution of copper sulfate ($CuSO_4$) as in the case of white phosphorus. These compounds, being sensitive to air, should be destroyed under an inert atmosphere (dry nitrogen).

---

* Pure phosphine does not ignite spontaneously and only burns in air at temperatures exceeding 150°C. However, the presence of traces (0.2%) of diphosphane (H_2P-PH_2) formed during the preparation of phosphine causes the mixture to ignite spontaneously in air.

2.21 Destruction of Isocyanates and Diisocyanates

Examples

Methyl isocyanate

$$CH_3 - N = C = O$$

Phenyl isocyanate

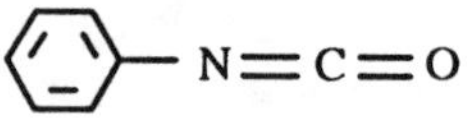

Di-isocyanatotoluene

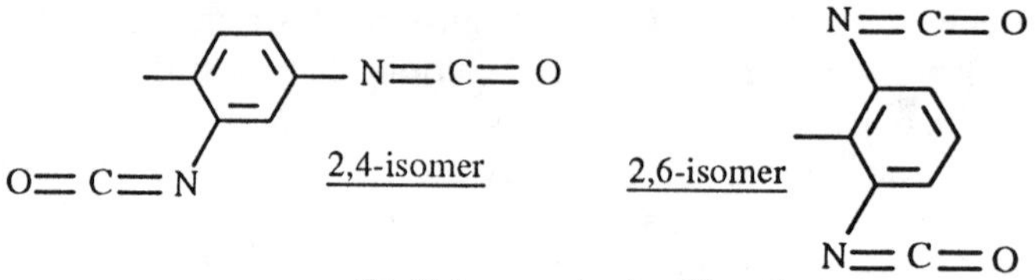

2,4-isomer

2,6-isomer

Bis(4-isocyanatophenyl)methane

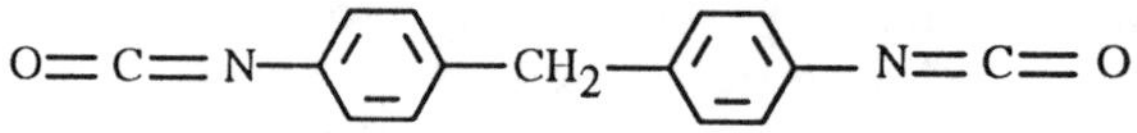

Principle of destruction

Organic isocyanates are highly reactive substances that are particularly sensitive to hydroxylated compounds: water, mineral bases, alcohols, and so on. Thus, in the presence of an alcohol, an isocyanate is transformed into a carbamate (urethane).

$$R_1 - N = C = O + R_2OH \longrightarrow R_1 - \overset{H}{\underset{}{N}} - \overset{O}{\underset{}{C}} - O - R_2$$

Organic isocyanates are destroyed by contact with water, mineral bases (e.g., sodium hydroxide, potassium hydroxide, ammonium hydroxide), or methanol. These reactions are exothermic, particularly in the case of aliphatic derivatives. Thus, with methyl isocyanate, attack by water is slow below 20°C, but may become very violent at higher temperatures in addition to provoking vaporization of more or less important quantities of nonreacted, toxic isocyanate. Base ($NaOH$, KOH, NH_4OH, etc.) or acid (H_2SO_4, etc.) catalysis accelerates this reaction.

Depending on their solubility, isocyanates may be destroyed by:

−addition to an ice-cold, 10% solution of ammonium hydroxide
−slow addition to a large excess of cold water
−slow addition to cold methanol

Special precautions

Protective gastight goggles
Rubber gloves

Safety shield

Alkyl isocyanates must be handled with great caution under a well-ventilated fume-hood. Isocyanate vapors may be absorbed using bubblers filled with 20% sodium hydroxide.

Remark

Organic isocyanates are strongly irritating to eyes (lachrimators), skin, and respiratory mucous membranes (acute bronchitis, acute pulmonary edema). They are also allergenic (asthma, etc.). Methyl isocyanate is particularly toxic and its inhalation may lead to rapid death (acute pulmonary edema).

Destruction using ammonium hydroxide

To a round-bottom flask equipped with a stirrer, a thermometer, and a condenser connected to a bubbler containing 20% sodium hydroxide, add a cold, 10% solution of ammonium hydroxide.

Slowly and cautiously add the isocyanate to be destroyed. Control the temperature of the reaction mixture and the rate of gas evolution. Allow to stir overnight. Store for disposal. This technique is particularly adapted to the destruction of small quantities of methyl isocyanate.

Destruction using an excess of methanol

This method is identical to the preceding one except that an excess of ice-cold methanol is used. This technique is particularly adapted to the destruction of diisocyanates.

2.22 Destruction of Mercaptans, Sulfides, and Other Sulfur Derivatives

Examples

Alkali sulfides
    Sodium sulfide
    Potassium sulfide
Mercaptans
    Methyl mercaptan (methanethiol)
    Ethyl mercaptan (ethanethiol)
    n-Butyl mercaptan (n-butylthiol)
Sulfides (thioethers)
    Dimethyl sulfide
    Diphenyl sulfide, etc.

Miscellaneous
  2-Mercaptoethanol
  Thioacetic acid
  Thiourea
  Thiram (tetramethylthiuram disulfide)
  Thiophen

Principle of destruction

Alkali sulfides (e.g., Na_2S, K_2S, etc.), hydrogen sulfide (H_2S), and organic derivatives in general (e.g., mercaptans, thioethers, disulfides, etc.) may be destroyed by oxidation in the presence of sodium hypochlorite.

$$Na_2S + 4\,NaOCl \longrightarrow Na_2SO_4 + 4\,NaCl$$

Organic sulfur derivatives (e.g., mercaptans, thioethers, carbon disulfide, etc.) may also be destroyed by incineration in an appropriate installation.

Hypochlorites: sodium hypochlorite (NaOCl) or calcium hypochlorite $Ca(OCl)_2$ oxidize sulfur compounds with formation of water-soluble derivatives. For example, thioethers are oxidized to sulfoxides and, more slowly, to sulfones.

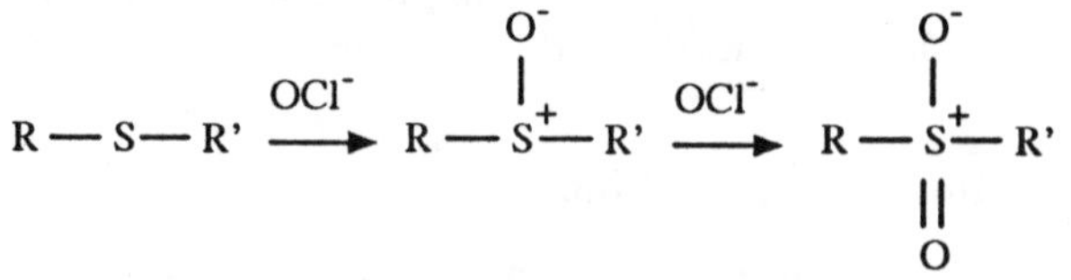

$$R-S-R' \xrightarrow{OCl^-} R-\overset{O^-}{\underset{}{\overset{|}{S}}{}^+-R'} \xrightarrow{OCl^-} R-\overset{O^-}{\underset{\overset{||}{O}}{\overset{|}{S}}{}^+-R'}$$

Similarly, mercaptans are oxidized to sulfonic acids.

$$R\text{-}S\text{-}H + 3\,NaOCl \longrightarrow R\text{-}SO_3H + 3\,NaCl$$

Hydrogen sulfide and mercaptans dissolve easily in alkaline media (NaOH, KOH) and these thiolate solutions may be easily destroyed by alkaline potassium permanganate which oxidizes these to alkali sulfonates.

$$R-SH \xrightarrow{KOH} R-S-K \xrightarrow{KMnO_4} R-SO_3K^+$$

The use of dimethylsulfoxide (DMSO) in numerous organic syntheses often leads to the formation of foul-smelling by-products which may be destroyed by treatment with an alkaline solution of potassium permanganate or sodium hypochlorite.

Special precautions

Protective goggles
Rubber gloves

If a sulfur-containing compound is released in the course of a reaction (e.g., hydrogen sulfide, mercaptan, etc.), trapping with an alkaline solution (such as

NaOH) or a 5% alkaline solution of potassium permanganate is necessary. It should be noted that low molecular weight mercaptans are of comparable toxicity to hydrogen sulfide, though their effects are less serious (headache, vomiting).

Elimination of spills

Cover the contaminated area with a large excess of a 10% solution of sodium hypochlorite (household bleach). Allow to stand, then wash the contaminated area with a strong soap solution containing bleach. Adding solid calcium hypochlorite directly to a sulfur derivative is not recommended; the exothermic reaction may lead to explosion or combustion of the mixture.

Destruction of mercaptans, thioethers, and so on

In a round-bottom flask equipped with a stirrer, a dropping funnel, and a thermometer, place:

–2,500 ml of 10% sodium hypochlorite (commercial solution)

While stirring at room temperature, slowly introduce the organic sulfur compound (0.5 mol). As it is oxidized, the sulfur compound dissolves and the temperature progressively rises.

Control the rate of addition of the sulfur derivative such that the temperature of the mixture is maintained at 45–50°C. If necessary, cool with an ice bath. Addition should be complete within 1 hour. Continue stirring for 2 hours, allowing the mixture to come to room temperature. In the case of mercaptans, the solution should be clear and odor-free. If not, oxidation should be continued by adding fresh sodium hypochlorite. With thioethers (R—S—R'), poorly soluble sulfones (R—SO_2—R') are formed which may precipitate. Filter and eliminate the precipitate in the same manner as other solid wastes. Insoluble organic sulfur derivatives may be dissolved with tetrahydrofuran.

Destruction of carbon disulfide

Carbon disulfide (CS_2) is a very dangerous substance; a volatile liquid (bp $= 46°C$), it spontaneously ignites at 100°C and is highly toxic (central and peripheral neurotoxicity). It may be destroyed by incineration after dilution in 10 parts of kerosene. Small quantities may be eliminated by oxidation with sodium hypochlorite.

$$S=C=S + 8\,NaOCl + 2\,H_2O \longrightarrow CO_2 + 2\,H_2SO_4 + 8\,NaCl$$

To 670 ml of 10% sodium hypochlorite (a 25% excess), slowly add, while stirring and while maintaining the temperature at 20–30°C, 3 ml of carbon disulfide (0.05 mol). Allow the mixture to stir 2 hours and store for disposal.

Destruction of alkaline sulfides

Alkaline sulfides (e.g., Na_2S, K_2S) may be destroyed in the same manner as mercaptans, by slow addition to an excess of sodium hypochlorite solution.

Destruction of hydrogen sulfide

Hydrogen sulfide should first be absorbed in a 10% solution of sodium hydroxide and this solution then treated with excess sodium hypochlorite or 10% potassium permanganate.

2.23 Destruction of Mercury and Its Derivatives

Examples

Elemental mercury	$Hg^{\bullet}$
Mercuric chloride	$HgCl_2$
Mercuric nitrate	$Hg(NO_3)_2$

Properties of mercury

A liquid at room temperature, elemental mercury, as well as some of its organic and inorganic derivatives, release significant quantities of vapor. These quantities increase with temperature. At 25°C, its saturated vapor concentration in air is of the order of $20\ mg/m^3$ or about 2.5 ppm. This ease of evaporation of mercury at ambient temperature may rapidly lead to dangerous air levels: the dispersion of $1\ cm^3$ of mercury in a medium-size laboratory may provoke toxic phenomena.

Mercury is very poorly soluble in water ($20–30\ \mu g/l$ between 20 and 30°C), in mineral bases, in dilute mineral acids (HCl, H_2SO_4, etc.), and in most organic solvents. It is soluble in hot, dilute nitric acid and in concentrated sulfuric acid. Mercury forms amalgams with numerous metals (e.g., Cu, Zn, Sn, Fe, Al, etc.) and reacts readily with certain elements (e.g., sulfur, iodine, etc.).

Principle of destruction

Although many methods have been proposed for the elimination of metallic mercury, their effectiveness seems quite variable.

Several techniques make use of the ease of formation of insoluble, nonvolatile mercuric sulfide when mercury comes in contact with sulfur or sulfur derivatives (e.g., sulfides, etc.).

$$Hg^{\circ} + S^{\circ} \longrightarrow HgS \downarrow$$

Other methods make use of the propensity of mercury to form solid amalgams with various metals (e.g., Cu, Zn, Sn, Fe). Mercuric sulfide (HgS) as well as amalgams may be easily recovered and eliminated with solid wastes. They should never be eliminated in the sink.

Mechanical methods (aspiration, etc.) or physicochemical methods allow direct recovery of metallic mercury. Recovered mercury must be stored in a hermetically sealed recipient.

Special precautions

Protective goggles
Rubber gloves
Mercury-specific protective respirator (if necessary)

Elimination of spills

Aspiration under vacuum.   Spilled mercury may be collected by aspirating into a vacuum flask with the help of a glass or steel tube attached to a vacuum pump protected with a cooled trap. The elimination of remaining mercury traces is effected using one of the techniques described later (transformation into zinc amalgam or a sulfide, etc.).

Freezing.   Mercury drops frozen by the addition of liquid nitrogen, dry ice, or a simple mixture of ice acetone, are easily recovered in solid form.

Transformation into mercuric sulfide

Treatment with calcium polysulfide.   Pour onto the contaminated surface (workbench, etc.) a 20% solution of calcium polysulfide. Allow to dry during 48 hours and avoid breathing the hydrogen sulfide vapors liberated. Sponge up the mixture with a moistened rag and combine with other solid wastes destined for destruction.

Treatment with a mixture of sulfur and lime.   Spread over the contaminated area (floor, etc.) a mixture composed of equal parts of sublimed sulfur and lime (hydrated calcium oxide). Allow 24 hours of contact. Add water to the mixture and collect by aspiration. Store in a tightly stoppered container before destroying with other solid wastes.

Treatment with sulfur.   Spread sulfur over the contaminated area. Allow to react overnight. Collect the mixture by aspiration with a vacuum pump. Store and eliminate with other solid wastes. Recent studies indicate this method is not as practical and efficient as other methods like the sponge, Baker mercury kits, and vacuum aspiration.

This procedure is also effective for trapping mercury vapors.

Transformation into solid amalgams

Copper amalgam.   Prepare a mixture of copper powder (nine parts) and potassium bisulfate (one part). Add enough water to make a smooth paste.

Apply this paste to the contaminated areas (joints and fissures on the workbench, etc.). Allow several days of contact to ensure complete reaction. Collect the amalgam (gray substance) and store before destroying.

A modification of this method utilizes a copper strip cut obliquely and preamalgamated on the oblique portion. The mercury is then collected on the tip and transferred to a recovery vial.

Remark

Zinc powder (moistened with a 0.1 N HCl solution) also forms an amalgam. The same is true of tin washed with a dilute solution of HCl. Aluminum powder also reacts to form an amalgam, but this procedure is not recommended.

Iron amalgam.   Into a polyethylene beaker equipped with a Teflon magnetic stirring bar, place a 1 N HCl solution and iron fragments. Slowly add the waste solids (mercury) or solutions containing water-soluble mercury salts [$HgCl_2$, $Hg(NO_3)_2$]. Allow to stir for 3 hours. Collect the iron amalgam by decantation and store in a tightly closed container before destroying.

Remark

The use of a "mercury sponge" also allows small quantities of spilled mercury to be collected. Saturate a household metallic sponge (copper or brass) with a mixture of zinc powder and 5% sulfamic acid and add a detergent dropwise. The surface to be decontaminated is wiped with the sponge and the collected mixture is stored in a container before being destroyed.

Mercury may also be directly recovered with a foam sponge of fine porosity (less than 0.5 mm). The mercury-soaked sponge is stored in a container before being destroyed.

The use of commercial kits

For cleaning small areas, decontaminating kits specific for mercury may be used. The kit manufactured by the J.T. Baker Company makes use of a mercury aspirator.

Absorption of mercury vapors

Mercury vapors emitted during the course of a reaction may be trapped by bubbling through a 20% solution of calcium polysulfide.

Mercury vapors expelled from a fume-hood may be trapped on active charcoal impregnated with iodine. Mercury so absorbed is released only at temperatures exceeding 160°C. These charcoal filters must be renewed frequently in order to maintain their effectiveness.

Preparation of iodine-impregnated active charcoal.   In a well-ventilated fume-hood, progressively mix 95 g of active charcoal and an ether solution of iodine (10 g in 30 ml of diethyl ether). A layer about 2 cm thick of the paste so obtained is spread on paper and allowed to dry. On direct contact, 1 g of active charcoal can absorb approximatively 1 ml of mercury. This mixture is stored in a container before being destroyed.

Open air storage of mercury

Although the best method of storing mercury is in a closed container, certain techniques require that mercury be left in contact with air (e.g., mercury joints, pressure boosters, etc.). In such cases, the protection provided by a layer of water or oil is of very limited effectiveness:

–liquid barriers (water, oil) are rapidly crossed by mercury vapors
–the rate of crossing depends on the thickness of the liquid layer
–paraffin oil is better than water at stopping the diffusion of mercury vapors

Under these circumstances, the best method consists of covering the mercury with a thick layer of water which in turn is covered with a film of paraffin oil. Mercury surfaces can also be covered with a layer of concentrated phosphoric acid. This has the added advantage of preventing oxidation.

2.24 Destruction of Metals and Their Derivatives

Examples

All metals except alkaline metals
Metal oxides
Metal salts
Organometallic compounds

Principal metals

Antimony	Chromium	Mercury	Tin
Arsenic	Cobalt	Nickel	Titanium
Barium	Copper	Osmium	Uranium
Beryllium	Iron	Silver	Vanadium
Bismuth	Lead	Strontium	Zinc
Cadmium	Manganese	Thallium	etc.

Many of these are heavy metals (Cd, Cr, Hg, Pb, Tl, etc.), and are consequently serious environmental pollutants.

Principle of destruction

In the metallic state (except for mercury, which is a liquid) or in insoluble forms (oxides, sulfides, etc.), metals may be directly eliminated as solid waste. In the form of water-soluble salts or of liquid-soluble organometallic derivatives, metals must be transformed into insoluble compounds which can be eliminated with solid wastes.

Since many metal derivatives are toxic to human beings and the environment, it is necessary to ensure that these are eliminated using techniques as nonpolluting as possible. Throwing metal derivatives down the drain should be avoided. In particular, water-soluble derivatives of heavy metals (e.g., Cd, Cr, Hg, Pb, Tl) should never be disposed of in this manner.

Two general techniques may be used to transform water-soluble metal salts into insoluble derivatives:

–transformation into insoluble hydroxides
–transformation into insoluble sulfides

In a few particular cases, the insoluble salts may be sulfates (e.g., Ba, Pb), carbonates (e.g., Ba, Pb, Sr), and so on.

Transformation into insoluble hydroxides

A large number of metals, in the presence of a strong mineral base (sodium hydroxide, potassium hydroxide, ammonium hydroxide), form insoluble metallic hydroxides. Some of these (e.g., Ag^+, Be^{++}, Cd^{++}, Cu^{++}, Ni^{++}, etc.) are insoluble in an excess of base.

Example: cadmium salts

$$Cd^{++} + 2\,OH^- \longrightarrow Cd(OH)_2 \downarrow$$

Many other metallic hydroxides (Cr^{++}, Pb^{++}, Sn^{++}, Zn^{++}, etc.) may be solubilized by an excess of base.

Example: chromium salts

$$Cr^{+++} + 3\,OH^- \longrightarrow Cr(OH)_3 \downarrow$$

$$Cr(OH)_3 + OH^- \longrightarrow CrO_2^- + 2\,H_2O$$

The quantitative precipitation of chromium salts (Cr^{+++}) as chromic hydroxide $Cr(OH)_3$ may be realized by their addition to a hot solution of an excess of ammonium chloride followed by addition of a sufficient quantity of ammonium hydroxide. Hexavalent chromium salts (chromates, dichromates) require a preliminary reduction to chromic salts (Cr^{+++}) with ferrous sulfate. For small quantities, heavy metals may be eliminated by filtration on a short, large diameter column filled with a mixture (1:2 to 2:1 parts) of granulated marble (Merck) and magnesium hydroxide or magnesium oxide.

Transformation into insoluble sulfides

Numerous water-soluble metallic salts form insoluble sulfides in the presence of hydrogen sulfide (H_2S) or of alkaline sulfides (e.g., NH_4HS, $(NH_4)_2S$, Na_2S, etc.). In practice, excess ammonium bisulfide (NH_4SH) in a strongly acidic medium (e.g., HCl, etc.) is used to precipitate numerous metals: Ag^+, Be^{++}, Cd^{++}, Hg^{++}, Os^{++++}, Pb^{++}, and so on.

Example: mercuric salts

$$Hg^{++} + NH_4HS \longrightarrow HgS\downarrow + NH_4^+ + H^+$$

In order to obtain as complete a precipitation as possible, it is necessary to work within a given pH range (Table 2.1). Thus, the formation of sulfides of certain metals (e.g., Co^{++}, Ni^{++}, Tl^+, Zn^{++}) requires a moderately acidic medium (pH $\cong$ 2–3) and precipitation by ammonium bisulfide must be performed in the presence of acetic acid or sodium acetate.

Example: precipitation of nickel salts

$$Ni^{++} + CH_3COO^- + NH_4HS \longrightarrow NiS\downarrow + CH_3COONH_4 + CH_3COOH$$

Arsenic and antimony do not precipitate in the presence of ammonium bisulfide and their precipitation in the form of sulfides must be effected with hydrogen sulfide in an acidic medium.

Sulfide (color)	Solubility	pH at start of precipitation	pH at end of precipitation
MnS (pink)	$10^{-9.6}$	6.7	8.2
MnS (green)	$10^{-12.6}$	5.2	6.7
FeS (black)	$10^{-17.2}$	2.9	4.4
Tl_2S	$10^{-20.3}$	2.05	5.05
NiS (black)	$10^{-18.5}$ to $10^{-25.7}$	2.25 to - 1.35	3.75 to -0.15
CoS (black)	10^{-20} to 10^{-25}	1.5 to - 1	3 to 0.5
ZnS	$10^{-21.6}$ to 10^{-24}	0.7 to - 0.5	2.2 to 1
CdS (red-brown)	$10^{-27.8}$	- 2.4	-0.9
PbS (black)	$10^{-27.9}$	- 2.45	-0.95
SnS (brown)	$10^{-25.0}$	- 1	0.5
Cu_2S	$10^{-47.6}$	- 11	-8
CuS	$10^{-35.2}$	- 6.1	-4.6
Ag_2S (black)	$10^{-49.2}$	- 12.8	-8.8
HgS (black)	10^{-52}	- 14.5	-13

Table 2.1 Properties of metallic sulfides.

Example: precipitation of arsenic salts

$$2\,As^{+++} + 3\,H_2S + H^+ \longrightarrow As_2S_3 + 6\,H^+$$

Other techniques of precipitation

Barium, an alkaline earth metal whose water-soluble salts are particularly toxic, may be rendered insoluble by addition of sodium carbonate, resulting in formation of barium carbonate.

Example: precipitation of barium salts

$$Ba^{++} + Na_2CO_3 \longrightarrow BaCO_3 + 2\,Na^+$$

Vanadium derivatives (e.g., V_2O_7, $VOCl_3$, Na_3VO_4, etc.) form poorly soluble ammonium vanadates in the presence of an ammonium salt (e.g., chloride, carbonate, etc.).

Example: precipitation of soluble derivatives of vanadium

$$V_2O_7 + 4\,NH_4^+ \longrightarrow 2\,NH_3 + H_2O + 2\,NH_4VO_3$$

Special precautions

Protective goggles
Rubber gloves
Safety shield (for particularly reactive compounds)

Precipitations using hydrogen sulfide must be performed under a properly functioning fume-hood. Because hydrogen sulfide and alkaline sulfides are foul-smelling, the trapping of emitted vapors by bubbling through a 10% sodium hydroxide solution is indispensable. The destruction can be completed by adding to a 10% potassium permanganate solution.

Remark

In addition to a potent irritating action on lungs (acute edema) and a nonnegligible neurotoxicity (cephalalgia, etc.), hydrogene sulfide is a dangerous cytotoxic substance (through inhibition of the cytochrome oxidase of the mitochondrial respiratory cycle).

Several insoluble sulfides of heavy metals (e.g., CdS, Ni_3S_2, etc.) and a few insoluble hydroxides ($Be(OH)_2$, etc.) are carcinogenic in animals and must be handled with appropriate care.

Antimony, arsenic, bismuth.   Antimony, arsenic, bismuth, and their derivatives, in the form of water-soluble chlorides, are precipitated by an excess of
hydrogen sulfide in a slightly acidic medium.

Precipitation technique.   Dissolve the metallic derivative in the minimum
amount of concentrated hydrochloric acid. Dilute with water until precipitation is complete (a white precipitate of SbOCl or BiOCl). Add a sufficient
quantity of 6 M HCl to dissolve this precipitate. Saturate the solution with a
stream of hydrogen sulfide. Filter the precipitate and wash to neutrality.
After drying, the sulfide so obtained may be destroyed with other metallic
wastes.

Silver, cadmium, mercury, osmium, lead.   The water-soluble nitrates of silver,
cadmium, mercury, osmium, lead, and their derivatives are quantitatively
precipitated by ammonium bisulfide (NH_4HS) in strongly acidic media
(pH $\cong$ 1) in the form of insoluble sulfides. Aqueous solutions of these compounds (water washes, etc.) must be concentrated before treatment in acid.

Precipitation technique.   This technique is particularly well suited to amalgams. In a porcelain vessel, dissolve the metal derivative (e.g., amalgam, salts,
oxides, etc.) in a minimum of concentrated nitric acid. Evaporate the solution
to dryness under a properly functioning fume-hood. Dissolve the residue in
500 ml of water and check that the pH is acid. Add a solution of ammonium
bisulfide dropwise until precipitation is complete. Allow to stand for several
hours. Filter, wash, and dry the precipitate. The metallic sulfide may be
eliminated with other solid wastes (except silver sulfide which may be recovered).

Barium and strontium.   Barium and strontium salts and oxides form insoluble
carbonates in the presence of sodium or ammonium carbonate.

Precipitation technique.   Dissolve the alkaline earth metal derivative in a
minimum of 6 N HCl. Filter. Neutralize the filtrate with 6 N NH_4OH and add
a 10% solution of sodium carbonate dropwise. Allow to stand. Filter, wash,
and dry the precipitate. The insoluble carbonate may be eliminated with other
solid wastes.

Beryllium, chromium.   Water-soluble derivatives of beryllium (Be^{++}) and of
chromium (Cr^{+++}) precipitate in the form of insoluble hydroxides when
treated with excess ammonia in an ammonium medium (i.e., ammonium
chloride).

   Chromium: optimum pH for precipitation = 9.5
   Beryllium: optimum pH for precipitation = 8.5

Precipitation technique.   Dissolve the metal derivative in a minimum of 6 N
HCl. Filter. Add solid ammonium chloride to the filtrate followed by 6 N
ammonia solution dropwise until the pH is distinctly alkaline (pH 9.5). Bring

to a boil and then allow to stand overnight. Filter and wash the precipitate to neutrality. After drying, the metallic hydroxide may be eliminated with other solid wastes.

Cobalt, tin, nickel, thallium.   Water-soluble forms of cobalt, tin, nickel, and thallium may be precipitated by ammonium bisulfide (NH_4HS) in a weakly acidic medium (acetic acid, sodium acetate, pH 3) in the form of insoluble sulfides.

Precipitation technique.   To an aqueous solution of the metallic derivative, acidified with acetic acid, add sodium acetate until the pH reaches 3–4. A solution of ammonium bisulfide (NH_4HS) is then added dropwise until precipitation is complete. After allowing to stand for several hours, the precipitate is filtered, washed to neutrality, and dried. The insoluble metallic sulfide may be eliminated with other solid wastes.

Remark.   Thallium sulfide (Tl_2S) is rapidly oxidized in air to soluble thallium sulfate. Thus, the separation must be performed quickly.

Vanadium.   To an aqueous solution of the vanadium derivative (e.g., Na_3VO_4, VCl_3, $VOCl_3$, $VOSO_4, 3H_2O$), add a solution of ammonia until an alkaline pH is reached. Bring to a boil and slowly add an excess of a 10% solution of ammonium chloride. Allow to stand. Filter and wash the precipitate with a saturated solution of ammonium chloride. After drying, the solid may be eliminated with other solid wastes.

2.25 Destruction of Alkaline Metals

Examples

Cesium
Lithium
Potassium
Rubidium
Sodium

Principle of destruction

The alkaline metals react violently with water, liberating hydrogen which may ignite. The reactivity of alkaline metals with water increases in the order $Li < Na < K < Ru < Cs$.

The reaction of lithium with cold water is moderate, but with hot water, the reaction is vigorous and the hydrogen liberated may ignite. With sodium, the action of water is not explosive below 40°C. In contrast, the highly exothermic reaction of water with potassium may be explosive even at 20°C and the hydrogen formed ignites spontaneously.

With rubidium and especially cesium, the reaction with cold water is very violent and leads to the immediate ignition of hydrogen. The reaction with simple alcohols (e.g., methanol, ethanol, isopropanol, etc.) is similar.

The destruction of alkaline metals is generally effected by contact with a higher alcohol (e.g., isopropanol, butanols, amyl alcohols, etc.).

Example: destruction of potassium in the presence of t-butanol

$$2\,(CH_3)_3C\text{-}OH + 2\,K \longrightarrow 2\,(CH_3)_3C\text{-}OK + H_2$$

Special precautions

Protective goggles
Rubber gloves
Polycarbonate safety shield

Work with dry equipment (i.e., scalpel, tweezers, etc.) and do not submit equipment for washing that may hold traces of alkaline metals.

Destruction of alkali metal residues

Lithium.   To a large vessel containing ice water, add the lithium in small portions. After complete reaction, store for disposal.

Lithium as well as calcium may also be destroyed by slowly adding them to a cold solution of dilute hydrochloric acid.

Sodium.   To a large vessel (e.g., steel cylinder) containing an excess of an alcohol (e.g., ethanol, isopropanol, n-butanol, t-butanol), add the sodium in small pieces.* Allow all the sodium to react before adding more. After complete reaction, neutralize the mixture with 10% HCl and store for disposal.

The drying of certain solvents (e.g., toluene, dioxane, etc.) with fresh sodium may lead to the formation of a protective coating of sodium hydroxide around the metal. Extreme care must be taken in destroying such residues. Sodium so protected reacts very slowly at first, but the reaction may become very violent or even explosive if the quantity of sodium is large.

Potassium, rubidium, cesium.   All manipulations should be performed behind a polycarbonate safety shield. To a dry, enameled vessel containing n-butanol or t-butanol, cautiously add small pieces of the metal residues.† Control the rate of addition such that all the alkaline metal is destroyed before an additional quantity is introduced. After complete reaction, and after having carefully verified that all traces of metal are absent, cautiously pour the mixture into excess water. Neutralize the solution with 10% HCl and store for disposal.

---

*The destruction of 1 g of sodium requires 13 ml of 95% ethanol.
†The destruction of 1 g of potassium requires 21 ml of t-butanol.

2.26 Destruction of Metal Carbonyls

Examples

Iron pentacarbonyl*	$Fe(CO)_5$
Nickel tetracarbonyl*	$Ni(CO)_4$
Zinc pentacarbonyl*	$Zn(CO)_5$

Principle of destruction

The metal atom of metal carbonyls is generally at a low level of oxidation (i.e., zero valency) and may be easily oxidized by various reagents (e.g., HNO_3, NaOCl, etc.). Thus, in the presence of dilute nitric acid, metal carbonyls are transformed by oxidation into the corresponding nitrates.

$$Ni(CO)_4 + 2\,HNO_3 \longrightarrow Ni(NO_3)_2 + 4\,CO + H_2$$

The reaction between nickel tetracarbonyl and nitric acid is very exothermic.

Oxidation may also be effected in the presence of excess (25%) sodium hypochlorite.

$$Ni(CO)_4 + NaOCl + H_2O \longrightarrow Ni(OH)_2 + NaCl + 4\,CO$$

Special precautions

Protective goggles
Rubber gloves
Polycarbonate safety shield

Work under a properly functioning fume-hood (keeping in mind that nickel carbonyl vapors are six times heavier than air). If necessary, use a gas mask specific for nickel carbonyl or a self-contained breathing apparatus.

Elimination of stocks

Eliminate stocks of metal carbonyls in dumping areas generally used for dangerous chemicals.

Destruction of small quantities of metal carbonyls

Oxidation in the presence of nitric acid. Under a properly functioning fume-hood, add to a round-bottom flask equipped with a stirrer, a thermometer, and a dropping funnel, a large excess of water at $+5°C$. Add the metal carbonyl to be destroyed and stir the mixture vigorously (metal carbonyls are

---

*Use of alloy manometers in the presence of carbon monoxide. L'Actualité Chimique. Safety note No. 5, pp. 67–68, March 1984.

carbonyl to be destroyed and stir the mixture vigorously (metal carbonyls are poorly soluble in water). Add dropwise a cold solution of 10% (v/v) nitric acid such that the temperature remains at $+5°C$. After completion of the addition, continue to stir the mixture for 1 hour while allowing it to come to room temperature. Pour the acid solution of metallic nitrate into a tin container destined for regulated waste disposal.

Oxidation in the presence of sodium hypochlorite. A solution of the metal carbonyl in tetrahydrofuran or other saturated hydrocarbon (at a concentration of approximately 5%) is carefully added while stirring and under a nitrogen atmosphere to a 10% solution of sodium hypochlorite (25% excess).

Remark

Nickel tetracarbonyl is a carcinogen and is very volatile (bp = 42°C), very flammable and, especially, very unstable liquid. It decomposes at 50°C and explodes at 60°C when heated in air. It reacts slowly with hydrochloric and sulfuric acids but violently with halogens and nitric acid.

Iron pentacarbonyl is a less reactive and more stable liquid than nickel tetracarbonyl. It can ignite in air. In the presence of light, it decomposes with liberation of carbon monoxide. It reacts with halogens and mineral acids.

All metal carbonyls are very toxic, especially nickel tetracarbonyl (100 times more toxic than carbon monoxide).

The products of destruction of metal carbonyls must never be eliminated down the sink (pollution by metals).

2.27 Destruction of Phosphorus and Its Derivatives

Examples

White phosphorus	P_4
Red phosphorus	P_4
Phosphine	PH_3
Phosphides	P^{-3}
Phosphorus trichloride	PCl_3
Phosphorus tribromide	PBr_3
Phosphorus pentachloride	PCl_5
Phosphorus oxychloride	$POCl_3$
Phosphorus pentaoxide	P_2O_5
Phosphorus pentasulfide	P_2S_5

Properties of phosphorus and its derivatives

Phosphorus (P_4).   Phosphorus exists principally in two forms:

—*White phosphorus*, very reactive (a potent irritant to skin and mucous membranes) and very toxic (hepato-, nephro-, and cardiotoxic). It may be contaminated by red phosphorus (i.e., yellow phosphorus). White phosphorus is very sensitive to oxidants. In powdered form, it is very pyrophoric. It burns in air with formation of irritating white fumes mainly composed of phosphorus pentaoxide.

$$P_4 + 5\,O_2 \longrightarrow P_4O_{10}$$

To limit oxidation, white phosphorus is stored under water.

—*Red phosphorus* is poorly reactive and practically nontoxic. In the presence of an alkaline solution, it liberates phosphine which may spontaneously ignite in air or explode.

$$P_4 + 3\,NaOH + 3\,H_2O \longrightarrow 3\,NaH_2PO_2 + PH_3$$

Phosphine (PH_3) and phosphides (P^{-3}).   Phosphine (PH_3) is a colorless, toxic (neurotoxic, more or less delayed acute pulmonary edema) gas with an obnoxious odor.

Phosphine generally contains traces of diphosphane (P_2H_4) which favors its ignition in air at room temperature, with emission of abundant white fumes of phosphorus pentaoxide.

$$4\,PH_3 + 8\,O_2 \longrightarrow P_4O_{10} + 6\,H_2O$$

Phosphides such as calcium phosphide (Ca_3P_2) zinc phosphide (Zn_3P_2), and aluminum phosphide (AlP), are hydrolyzed on contact with water, liberating phosphine.

$$AlP + 3\,H_2O \longrightarrow Al(OH)_3 + PH_3 \nearrow$$

Phosphorus halides and oxyhalides.   Phosphorus halides (e.g., PCl_3, PBr_3, PCl_5) and oxyhalides (e.g., $POCl_3$, $POBr_3$, etc.) are very reactive substances which are violently hydrolyzed upon contact with water.

$$POCl_3 + 3\,H_2O \longrightarrow 3\,HCl + H_3PO_4$$

Phosphorous
oxychloride

The hydrolysis of phosphorus trichloride (PCl_5) leads to formation of phosphorus acid (H_3PO_3) which can decompose, releasing spontaneously flammable phosphine (PH_3).

$$PCl_3 + 3 H_2O \longrightarrow H_3PO_3 + 3 HCl$$

$$4 H_3PO_3 \longrightarrow 3 H_3PO_4 + PH_3$$

Phosphorus acid   Phosphoric acid

The hydrolysis of phosphorus pentachloride (PCl_5) leads directly to phosphoric acid.

$$PCl_5 + 4 H_2O \longrightarrow H_3PO_4 + 5 HCl$$

All phosphorus halides and oxyhalides are skin and mucosa (e.g., eyes, lungs, etc.) irritants.

Phosphoric oxide (P_2O_5). Phosphoric oxide (or phosphorus pentaoxide), used mainly as a dehydrating agent, exists in dimeric form (P_4O_{10}). It reacts violently with water, releasing heat and principally forming phosphoric acid.

$$P_4O_{10} + 6 H_2O \longrightarrow 4 H_3PO_4$$

Phosphorus pentasulfide (P_2S_5). Phosphorus pentasulfide is hydrolyzed in water, forming hydrogen sulfide, a very foul-smelling and toxic gas.

$$P_2S_5 + 8 H_2O \longrightarrow 5 H_2S + 2 H_3PO_4$$

Principal methods of destruction

Phosphorus, phosphine, phosphides. Phosphorus, phosphine, and phosphides may be destroyed by oxidation with copper sulfate. The reaction with white phosphorus itself is complex and leads to formation of phosphoric acid.

$$P_4 + 10 CuSO_4 + 16 H_2O \longrightarrow 4 H_3PO_4 + 10 Cu° + 10 H_2SO_4$$

Phosphine may be destroyed by oxidation in the presence of copper sulfate, but in order to prevent an explosive reaction in the presence of air, destruction should be conducted under a nitrogen atmosphere.

Phosphides are easily eliminated by oxidation in the presence of excess sodium hypochlorite.

Red phosphorus may be destroyed by potassium chlorate.

$$6 P + 5 KClO_3 + 9 H_2O \longrightarrow 6 H_3PO_4 + 5 KCl$$

Phosphorus halides, oxyhalides, phosphoric oxide, phosphorus pentasulfide. Phosphorus halides, oxyhalides, phosphoric oxide, and phosphorus pentasulfide may all be destroyed by treatment with ice-cold water or, better, with an alkaline solution (e.g., NaOH, Na_2CO_3, $NaHCO_3$, $Ca(OH)_2$, NH_4OH, etc.).

$$PCl_5 + 4 NaOH \longrightarrow Na_3PO_4 + 5 HCl$$

Special precautions

Protective goggles
Rubber gloves

Approved respirator (specific for phosphine, etc.)
Safety shield (for very reactive compounds)

For the destruction of phosphorus compounds likely to release phosphine, all operations should be conducted under a properly functioning fume-hood.

Destruction of white phosphorus

Under a properly functioning fume-hood, place 800 ml (0.8 mol) of a 1 M solution of copper sulfate in a 2 l beaker equipped with an efficient stirrer. Add 5 g (0.16 mol) of phosphorus in small pieces and allow the mixture to stand for 1 week. The phosphorus slowly disappears and a light, black precipitate of copper and copper phosphide (Cu_3P_2) forms. The reaction is finished when all the white phosphorus has disappeared. While remaining under the fume-hood, the precipitate is collected by filtration and transferred to a beaker equipped with a stirrer and containing 500 ml of 10% sodium hypochlorite. Stirring is maintained for 1 hour in order to totally oxidize the phosphide into phosphate. The resulting copper phosphate is eliminated as the insoluble sulfide together with other solid wastes.

Phosphine and phosphides are destroyed in the same manner as white phosphorus.

Destruction of red phosphorus

To a round-bottom flask equipped with a condenser and an efficient stirrer, add 2 l of 1 N H_2SO_4, 33 g (0.27 mol, 100% excess) of potassium chlorate followed by 5 g (0.16 mol) of red phosphorus in small portions.

Reflux the mixture until all the phosphorus has dissolved (usually 5–10 hours depending on the size of the phosphorus pieces). Cool the mixture to room temperature and destroy the excess chlorate by adding 14 g of sodium bisulfite. The solution may then be stored for proper waste disposal.

Destruction of liquid phosphorus halides and oxyhalides

This technique is applicable to phosphorus trihalides (e.g., PCl_3, PBr_3, etc.) and to phosphorus oxyhalides (e.g., $POCl_3$, $POBr_3$, etc.).

Under a properly functioning fume-hood, add 600 ml of 2.5 N NaOH (1.5 mol) to a 2 l round-bottom flask equipped with a stirrer, a dropping funnel, and a thermometer. Cool the solution to $+5°C$ and carefully add dropwise the phosphorus derivative (0.5 mol) to be destroyed. Control the temperature by the addition of ice. Allow the solution to warm to room temperature. After neutralization (pH 7), the solution may be stored for proper waste disposal.

Destruction of phosphorus pentachloride

Under a properly functioning fume-hood, add a solution of 5 N NaOH (50% excess) to a beaker equipped with a stirrer and a thermometer. Cool the solution to +5°C by the addition of ice before adding the phosphorus pentachloride in small portions. When all has dissolved, treat as described previously.

Destruction of phosphoric oxide

Under a properly functioning fume-hood, operate as for phosphorus penta-chloride using only ice water. The reaction is very exothermic and releases irritating, toxic vapors. Old stocks of phosphoric oxide used as dehydrating agent should be carefully destroyed after first dissolving the protective surface layer (a mixture of phosphoric and polyphosphoric acids) with ice water.

Destruction of phosphorus pentasulfide

Phosphorus pentasulfide (P_2S_5) may be destroyed by slow addition to an ammonia solution (diluted by half) held at +5°C. The addition of an aqueous solution of $KMnO_4$ allows destruction of the hydrogen sulfide released in the process.

2.28 Destruction of Peroxide-Forming Substances

Examples

Numerous products (e.g., solvents, monomers, alkaline metals, and derivatives, etc.) fix oxygen in the presence of air, forming peroxide derivatives which are often very unstable (Table 2.2).

Acetylenes (methyl acetylene, etc.)
Alkoxides (sodium isopropoxide, etc.)
Aldehydes (acetaldehyde, cinnamaldehyde, etc.)
Alkaline amides (sodium amide, etc.)
Allylic compounds (ethyl allyl ether, etc.)
Benzylic compounds (cumene, etc.)
Organometallics (magnesium, zinc, etc.)
Vinylic compounds (vinyl acetate, etc.)
Cyclenes (cyclohexene, pinenes, limonene, etc.)
Dienes (1,3-butadiene, etc.)
Dihydropyran and tetrahydropyran ethers, etc.
Ethers (diethyl ether, diisopropyl ether, etc.)
Haloalkanes (vinylidene chloride, etc.)
Potassium, etc.

Peroxide-forming mineral compounds are generally reactive (e.g., alkaline metals, alkoxides, amides, organometallic derivatives, etc.). Peroxide-forming organic compounds possess, as a general rule, a hydrogen atom which is easily activated in the presence of oxygen (i.e., free radical reactions) (Table 2.2).

Functional group	Type of hydrogen	Peroxide-forming compounds	
$R_3-\!\!\!\underset{R_2}{\overset{R_1}{\mid}}\!\!\!-H$	Tertiary hydrogen	Methyl cyclopentane Decalin Chloroform	
$R-\!\!\!\underset{CH_3}{\overset{CH_3}{\mid}}\!\!\!-H$	Isopropyl hydrogen $R = $ phenyl $R = $ OH $R = CH_2OH$ $R = $ C-CH$_3$ $\quad\ \ \overset{\|}{\underset{O}{}}$	Cumene Isopropyl alcohol Isobutyl alcohol Isobutyl methyl ketone	
$\underset{R_5}{\overset{R_4}{}}\!\!>\!\!=\!\!<\!\!\underset{\underset{H}{R_2}}{\overset{R_3}{\underset{\mid}{}}}\!\!R_1$	Allylic hydrogen	Cyclohexene Cyclooctene	
benzene ring $-\!\!\!\underset{R_2}{\overset{R_1}{\mid}}\!\!\!H$	Benzylic hydrogen	Tetralin Cumene	
$R_1-\!\!\!\underset{H}{\overset{Z-R_3}{\mid}}\!\!\!-R_2$	Hydrogen α to a heteroatom $Z = O$	Ethers Diethyl ether Diisopropyl ether Monoglyme Diglyme	Tetrahydrofuran Dioxane Cellosolves... Hemiketals, Ketals Acetal
	$Z = \overset{H}{\underset{\mid}{N}}-R_3$	Secondary amines Pyrrolidine Piperidine	
	$Z = \overset{R_3}{\underset{\mid}{N}}\!\!-\!\!\overset{O}{\underset{R_4}{\diagdown\!\!\diagup}}$	Amides Lactams Ureas	
$\underset{H}{\overset{R}{}}\!\!>\!\!=\!\!O$	Aldehyde hydrogen	Acetaldehyde Benzaldehyde	
$\underset{R_3}{\overset{R_2}{}}\!\!>\!\!=\!\!<\!\!\underset{H}{\overset{R_1}{}}$	Vinylic hydrogen	Styrene Vinyl acetylene Vinyl pyridine Vinyl chloride Vinylidene chloride Trichloroethylene	Vinyl ethers Vinyl esters Acrylates Methacrylates Butadiene Chloroprene
$R-\!\!\!\equiv\!\!\!-H$	Acetylenic hydrogen	Propyne Diacetylene	

Table 2.2   Principal peroxide-forming compounds.

Formation of peroxide derivatives

The formation of peroxide derivatives corresponds to an autoxidation reaction of the free radical type involving the presence of molecular oxygen and often of initiators which serve as catalysts (e.g., trace metals such as Fe, Cu, Mn, Co, peroxides, light, heat, etc.).

The accumulation of peroxides in a product depends on the difference between their rate of formation and the kinetics of their decomposition. Thus, certain highly reactive compounds such as organometallics are a source of peroxide accumulation at low temperature since their rate of decomposition is lower than that of their formation.

As a general rule, it is preferable to store peroxide-forming compounds away from heat and light (sunlight).

However, volatile solvents likely to form peroxides, such as highly flammable ethers, should never be stored in a refrigerator unless spark-proof. For instance, diethyl ether, whose flash point is at $-45°C$, must never be kept under these conditions.

Bottles containing peroxide-forming substances should be labeled such that their dates of receipt and opening are indicated.

Very often, the dangers presented by peroxide-forming liquid substances, though well known, are underestimated, resulting in frequently observed explosions during the course of their distillation. Such explosions, often of great force, may also occur when a recipient containing such highly peroxidized substances are simply moved. Similarly, in attempting to open a bottle, the peroxides localized near the cap may spontaneously explode due to friction.

Peroxide-forming substances utilized after a long storage period (1–12 months depending on the product) must imperatively be checked for the possible presence of peroxides.

If peroxides are present, the product must either be treated to eliminate the peroxides or be destroyed (preferably by incineration). Depending on the way they are stored, the various peroxide-forming substances may be classed into three categories: A, B, and C (Table 2.3).

Category A: Compounds presenting a risk of peroxidation during storage

For category A, a test for the presence of peroxides must necessarily be effected at least every 3 months after the container is first opened. For example, diisopropyl ether, the solvent which is most easily peroxidized, sometimes explodes simply as a result of handling aged lots of this ether. Diisopropyl ether is first autoxidized to a relatively stable dihydroperoxide which, with time, is transformed into cyclic peroxides of acetone, a dimer (F = 132°C) and a trimer (F = 98°C), both very explosive and probably responsible for the explosions observed in the course of handling aged batches of this ether. As a comparison, the sensitivity to shock of these two acetone peroxides exceeds that of mercury fulminate. These two peroxides do not cause iodine to be released from potassium iodide (the peroxide test) and are resistant to most of the agents known to destroy peroxides (except combustion, of course). It should be noted that the autoxidation of diisopropyl ether occurs as easily in the absence as in the presence of light.

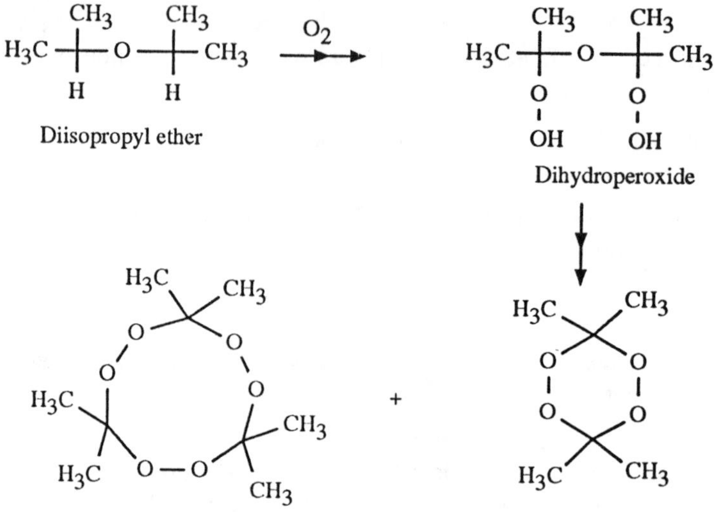

Diisopropyl ether

Dihydroperoxide

Trimer   Dimer

Easily peroxidized ethers such as diisopropyl ether may be advantageously replaced by methyl *t*-butyl ether (MTBE). This ether, with a low tendency to peroxidation (bp = 55°C), has found, as a result, numerous applications as a solvent for extraction and chromatography.

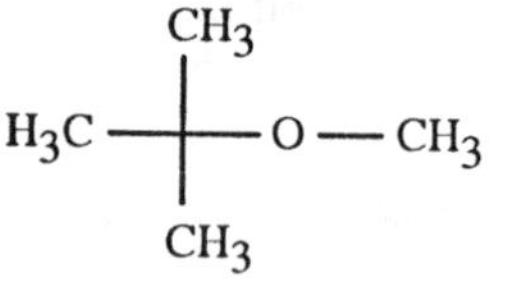

methyl t-butyl ether (MTBE)

Among the inorganic substances, potassium, during the course of storage, may undergo slow auto-oxidation in air (even when covered with mineral oil), becoming coated with a yellowish solid containing potassium superoxide (KO_2). By cutting "aged" potassium with a knife, small sparks may be produced, setting off an explosion. The latter is apparently due to the exothermic reaction between the superoxide and the freshly cut potassium. It is thus ill-advised to cut potassium that is covered by a protective yellowish solid, a sign that superficial autoxidation has occurred. Such peroxidized potassium must in this case be destroyed by slowly adding it to excess *n*-butanol or *t*-butanol.

A Risk of Peroxidation during Storage	B Risk of Peroxidation upon Concentration	C Risk of Polymerization by Peroxidation
Diisopropyl ether Divinyl acetylene Potassium Sodium amide	Diethyl ether Tetrahydrofuran Dioxane Monoglyme[1] Diglyme[2] Cellosolve[3] Isopropyl alcohol Isobutyl alcohol Isobutyl methyl ketone Methyl cyclopentane Cyclohexene Cumene Tetralin Decalin	Styrene Butadiene Vinyl acetate Vinyl chloride Vinyl pyridine Chloroprene

[1]monoglyme = $CH_3OCH_2CH_2OCH_3$, 1,2-dimethoxyethane, the dimethyl ether of ethylene glycol.
[2]diglyme = $CH_3OCH_2CH_2OCH_2CH_2OCH_3$, bis(2-methoxyethyl) ether, the dimethyl ether of diethylene glycol.
[3]cellosolve = $CH_3CH_2OCH_2CH_2OH$, 2-ethoxyethanol, the ethyl ether of ethylene glycol.

Table 2.3   Principal compounds likely to form peroxides during storage.

$$KO_2 + 3 K \longrightarrow 2 K_2O$$

Category B: Compounds presenting a risk of peroxidation as a result of concentrating their solutions

Compounds in category B, for the most part solvents, must not be stored more than 12 months after a container is st opened, unless an appropriate test indicates that peroxides have not accumulated.

In general, peroxides decompose in an explosive manner between 60 and 140°C. Because of this, the distillation of peroxide-forming liquids is dangerous if the distillation residue containing peroxide derivatives is heated above their decomposition temperature. These solvents must be distilled under an inert atmosphere (nitrogen or argon) and it is recommended to leave at least 10% of the initial volume as residue (20% for diethyl ether and diisopropyl ether). Freshly distilled, these solvents rapidly form more peroxides unless they are stabilized with an antioxidant. Thus, with tetrahydrofuran, the formation of peroxides begins on the third day after distillation, while with diethyl ether, peroxides appear on the eighth day.

It is generally preferable to store these solvents after distillation in containers protected from light (e.g., wrapped with aluminum foil), at a moderate temperature, in the presence of a peroxidation inhibitor (e.g., molecular sieves), and, if possible, under an inert atmosphere (i.e., nitrogen or argon).

Category C: Compounds presenting a risk of polymerization by peroxidation

In category C are found monomers that cannot be conserved more than 1 year unless no peroxides have accumulated (negative test). These monomers are generally supplied with a stabilizing antioxidant (e.g., phenols, amines).

It is usually preferable to store these compounds in small quantities, at low temperature, and under an inert atmosphere (i.e., nitrogen or argon). These monomers must only be distilled under an inert atmosphere after addition of a polymerization inhibitor (e.g., hydroquinone, etc.).

Destruction

In most cases, the peroxidation of organic compounds having an active hydrogen (a hydrogen α to an ether function, etc.) first leads to formation of a hydroperoxide (detectable by the potassium iodide test) which may then evolve either to a peroxide (undetectable by the potassium iodide test) or to a peroxidized polymer (as, e.g., in the peroxidation of diisopropyl ether).

While hydroperoxides are often easily destroyed by reduction, this is not so for peroxides and polymers which require much more drastic techniques. These compounds may be destroyed by more powerful reducing agents, for example, triphenylphosphine $[(C_6H_5)_3P)]$, lithium aluminum hydride $(LiAlH_4)$, or ferrous sulfate in sulfuric acid. Various types of compounds may be employed to destroy peroxide derivatives (Table 2.4).

The choice of the destruction method depends on the nature of product to be de-peroxidized as well as its quantity. The efficiency of the different types of destructive agents is variable, often depending on the operating conditions. Only the most commonly used laboratory techniques, mainly used to remove peroxides from solvents, will be discussed.

Distillation over sodium.   This is the oldest technique, requiring, to be effective, pretreatment with potassium hydroxide pellets. This technique may be applied to ethylene derivatives (e.g., cyclohexene, etc.) and to certain ethers (e.g., THF, dioxane, etc.). It permits elimination of only those hydroperoxides which form water-soluble alkali salts and must be performed cautiously. The direct distillation of THF in the presence of potassium hydroxide pellets may provoke a violent explosion. With ethers that are poorly soluble in water, the pretreatment may be effected by vigorous stirring with a concentrated solution of sodium or potassium hydroxide (e.g., stirring for 1.5 hours 10 parts of diethyl ether with 1 part of 23% aqueous sodium hydroxide). After drying over anhydrous calcium chloride, the ether may be distilled from sodium under an inert atmosphere. The addition of benzophenone to the distillation flask allows

Alkaline metals and derivatives	sodium sodium + ethanol sodium hydroxide potassium hydroxide
Metal derivatives	cuprous chloride stannous chloride tin + hydrochloric acid cerium hydroxide ferrous sulfate zinc + hydrochloric acid
Mixed hydrides	lithium aluminum hydride sodium borohydride
Sulfur derivatives	sodium bisulfite sodium sulfite mercaptans
Tertiary phosphines	triphenyl phosphine
Miscellaneous	activated alumina ion exchange resins (Dowex 1)

Table 2.4   The types of compounds used to destroy peroxide derivatives.

the extent of water and peroxide elimination to be monitored; a strongly colored (blue-violet) ketyl complex is formed when these substances are no longer present in solution.

The solvents so freed from peroxides may be stored over freshly cut sodium (e.g., wire, etc.) since this substance inhibits the formation of peroxides. It is, however, preferable to use less hazardous molecular sieves for this purpose.

Distillation over stannous chloride.   Stannous chloride is an excellent agent for the reduction of the peroxides formed in secondary alcohols (e.g., isopropyl alcohol, isobutyl alcohol, etc.) and in ethers (e.g., THF, dioxane, cellosolves, etc.). For example, 1 l of isopropanol may simply be left in contact with 10–15 g of stannous chloride for a few hours to ensure elimination of peroxides. However, if the peroxide test (potassium iodide) is still positive after this time, 5 g of fresh stannous chloride may be added and the mixture refluxed for 30 minutes. After addition of 200 g of calcium oxide (lime), the mixture is refluxed for 4 hours and then distilled. The distilled isopropanol is best stored over molecular sieves (5 Å).

As with most peroxide-forming solvents, peroxides appear again after a few days of storage.

Treatment with ferrous sulfate.   Ferrous salts (Fe^{++}) in solution are very effective in destroying hydroperoxides. The use of solid ferrous sulfate ($FeSO_4$), is not recommended, however, due to its limited effectiveness. Moreover, the distillation of THF from solid ferrous sulfate has led to explosions.

$$R\text{-}O\text{-}O\text{-}H + 2\,Fe^{++} + 2\,H^+ \longrightarrow ROH + 2\,Fe^{+++} + H_2O$$

The use of aqueous solutions of ferrous sulfate in acidic media is reserved for ethers which are poorly soluble in water (this method is thus unsuitable for THF, dioxane, and glymes). In addition, certain cyclic peroxides such as those of acetone formed during the course of diisopropyl ether autoxidation are resistant to such treatment. In this case, vigorous stirring with 50% sulfuric acid is necessary.

The reduction of peroxide derivatives, and especially of hydroperoxides, by ferrous sulfate is generally rapid and very exothermic. It is necessary in some cases to pre-cool the ether (e.g., diethyl ether) in order to prevent its boiling during treatment.

Example.   Into a 3 l separatory funnel containing 1 l of cold diethyl ether, slowly add 10–20 ml of an acidic ferrous sulfate solution (60 g of ferrous sulfate +6 g c—H_2SO_4 + 110 ml of water). When all the green ferrous sulfate turns brown, then it has been completely reduced. It is thus necessary to repeat the operation until a negative peroxide test is obtained in the ether phase (potassium iodide test). After washing with water to neutrality, the ether is dried over calcium chloride (100–200 g) for at least 24 hours. The ether is then distilled under an inert atmosphere and stored over molecular sieves, protected from air.

Distillation from lithium aluminum hydride.   Lithium aluminum hydride ($LiAlH_4$), a powerful reducing agent, is very sensitive to water and to oxygen and it may be used for the purification of certain ether solvents (e.g., THF, dioxane, diglyme, etc.), however, this operation presents numerous risks:

–the introduction of powdered lithium aluminum hydride into the sol-vent-containing flask may cause the mixture to catch fire,
–the addition of solvent to solid lithium aluminum hydride may provoke an explosion, even under an inert atmosphere,
–the introduction of air into a distillation flask at the end of the operation, may lead to extremely violent reactions (e.g., explosion, combustion), especially if the distillation residue is still warm.

Treatment with sodium sulfite.   In alkaline media (pH 9.5), sodium sulfite ($Na_2SO_3, 7H_2O$) is a good reducer of hydroperoxides. Sodium bisulfite ($NaHSO_3$) and sodium hyposulfite ($Na_2S_2O_7$) are less effective.

$$R\text{-}O\text{-}O\text{-}H\text{-} + Na_2SO_3 \longrightarrow R\text{-}OH + Na_2SO_4$$

Treatment with potassium iodide.

- Benzoyl peroxide. Peranhydrides poorly soluble in water such as benzoyl peroxide may be destroyed by the action of an alkali iodide (NaI or KI) in acetic acid.

$$R\overset{O}{\overset{\|}{-}}\!O - O \overset{O}{\overset{\|}{-}}R\ +2\,KI\ \longrightarrow\ 2\,R\overset{O}{\overset{\|}{-}}\!O-K\ +I_2$$

To a solution of 70 ml of acetic acid containing 0.022 mol (10% excess) of potassium iodide, slowly add, with stirring and at room temperature, 0.01 mol of peranhydride. The solution rapidly turns brown due to the formation of iodine. Allow to stir for 1 hour. Treat the solution with an excess of sodium hyposulfite such that the solution turns colorless and then store for proper waste disposal.

- Dialkyl peroxides. Most dialkyl peroxides (R—O—O—R') are not destroyed at room temperature by reagents which decompose peranhydrides (e.g., ferrous sulfate, iodides in acetic acid, or ammonium hydroxide). They may be destroyed by the preceding technique by adding 1 ml of 36% HCl. The solution is then heated to 90–100°C for 30 minutes and then left at room temperature for 5 hours. The resulting brown solution is treated as with benzoyl peroxide.

Treatment with active alumina. Activated, basic alumina (basic aluminum oxide), in addition to its dehydrating properties, allows elimination of peroxide derivatives. Its capacity for elimination depends on the water content of the solvent to be treated. This technique is effective both for water-soluble solvents (except secondary alcohols) and those that are only slightly soluble in water.

The quantity of alumina necessary depends on the nature of the solvent and the quantity of peroxide derivatives to be eliminated. As an example, percolation through a column containing 30 g of alumina (activated basic aluminum oxide 90 from Merck, e.g.) allows 250 ml of diethyl ether (or 100 ml of diisopropyl ether or 25 ml of dioxane) to be de-peroxidized.

While hydroperoxides (R—O—OH) are apparently well retained by alumina, this is not so for peroxides (R—O—O—R').

It is recommended that, after this operation has been completed, the alumina columns should not be simply put aside; the nondegraded peroxide derivatives may in fact concentrate at the top of the absorbant and cause the column to explode. The used alumina must be carefully eliminated by immersing it in a large volume of water.

This very simple technique of peroxide elimination may be used for the majority of ethers, for tetralin, and for decalin.

Contact with molecular sieves. Peroxides may be eliminated from ethers by percolation through a column containing type 13X (10 Å) molecular sieves. As with alumina, these sieves only retain peroxide derivatives and thus cannot be reactivated without a risk of explosion. Reactivation is possible, however, if 4 Å (4–8 mesh) molecular sieves impregnated with a color indicator (Baker, Sigma, or Merck) are used. It is probable that, in this case, the color indicator is composed of cobalt salts that ensure the reduction of the hydroperoxides.

$$R\text{-O-OH} + Co^{++} \longrightarrow R\text{-O}^- + OH^- + Co^{+++}$$

This technique is particularly effective for diethyl ether and diisopropyl ether, but less so for THF. It is useless for dioxane (which mainly contains cyclic peroxides).

Example.   Under a nitrogen atmosphere, 5% by weight of 4 Å molecular sieves are added to the peroxide-containing solvent and the mixture is left standing several days until a negative hydroperoxide test (potassium iodide) is obtained. With THF, it is preferable to reflux for 4 hours in the presence of 5% by weight of 4 Å molecular sieves.

2.29 Destruction of Selenium and Tellurium and Their Derivatives

Examples

Selenium dioxide	SeO_2
Sodium selenite	Na_2SeO_3
Selenium tetrachloride	$SeCl_4$
Phenylselenenic acid	$C_6H_5SeO_2H$
Phenylselenenic anhydride	$(C_6H_5SeO)_2O$
Diphenyl diselenide	$(C_6H_5Se)_2$
Tellurium dioxide	TeO_2
Potassium tellurate	K_2TeO_3

Principle of destruction

Inorganic selenium derivatives.   In strong hydrochloric acid (9 N) and in the presence of sulfurous anhydride (SO_2) or of sodium sulfite (Na_2SO_3), inorganic selenium derivatives precipitate in the form of red selenium, often colloidal and which, upon heating, becomes gray-black. The latter can then be eliminated by filtration on celite.

$$SeO_2 + SO_2 + 2H^+ \longrightarrow Se\downarrow + H_2SO_4$$

Inorganic tellurium derivatives.   In the case of inorganic tellurium derivatives, precipitation in the form of black tellurium is effected in the presence of sulfurous anhydride or of sodium sulfite in more dilute hydrochloric acid (3 N).

Organic selenium derivatives.   Organic selenium derivatives (e.g., phenylselenenic acid, phenylselenenic anhydride, phenylselenenic halides, etc.) are first oxidized with dilute nitric acid and then treated with an excess of sodium sulfite, transforming them into selenium.

Special precautions

Protective goggles
Rubber gloves
Operate under a properly functioning fume-hood

Remark

Selenium is considered only a weakly toxic element. Some of its inorganic derivatives (e.g., SeO_2, $SeOCl_2$, $SeOBr_2$, etc.) are powerful skin, eye, and lung irritants (acute edema). Hydrogen selenide (H_2Se), a highly irritating and hepatotoxic foul-smelling gas, is the most toxic derivative. The toxicity of organic selenium derivatives is not well known but appears to be greater in the case of the more volatile compounds.

Destruction

Techniques of selenium precipitation.
* Inorganic derivatives.   The selenium derivative is dissolved in 9 N HCl. To the cooled solution, add, while stirring, a saturated aqueous solution of sulfurous anhydride or a solution of sodium sulfite in excess. The solution is brought to a boil in order to convert the red selenium into black selenium. Allow to stand and filter on celite. Eliminate with other solid wastes.

* Organic derivatives.   Oxygenated organic derivatives (e.g., phenylselenenic acid, phenylselenenic anhydride, etc.) are dissolved in 10% nitric acid. The solution is brought to a boil under a properly functioning fume-hood (nitrogen oxides are released). The solution is cooled before adding c—HCl such that a 2–4% concentration is obtained. An excess of a saturated sodium sulfite solution is then introduced slowly while stirring.

2.30 Destruction of Carcinogenic Substances

Substances having carcinogenic properties may be found in numerous chemical families (e.g., metals, aromatic hydrocarbons, derivatives containing halogens, nitrogen, oxygen, sulfur, phosphorus) and no general method exists for their destruction. For large quantities of organic compounds, high-temperature incineration (100°C or higher for some halogenated derivatives) in special furnaces is the method of choice. Heavy metal derivatives (e.g., salts and oxides of Cd, Ni, Cr, Pb, etc.) may be transformed into insoluble sulfides, but some metallic sulfides (e.g., Ni_3S_2, CdS) are themselves endowed with carcinogenic properties.

Carcinogenic substances destined for destruction may be either in a pure state (gases, solids, or liquids) or as part of a mixture. Generally, they are in aqueous or organic (oil or solvent) solutions. They may also contaminate laboratory equipment, workbenches, or even the floor. In biological laboratories, carcinogenic substances may also be found in the materials used for animal experiments (e.g., the animals themselves, cages, litter, etc.). The techniques used to eliminate these substances depend on the type of material to be destroyed or decontaminated.

The criteria most often taken into consideration in choosing a method of destruction are:

–the possibility of operating easily on a laboratory scale,
–a moderate cost,
–high effectiveness; in particular, the final residues of destruction must not themselves be mutagenic (as in the case of aflatoxins treated only with sodium hypochlorite).

As a general rule, techniques that make use of toxic or polluting reagents (e.g., hexavalent chromium salts, etc.) should not be employed. The techniques most often retained for the destruction of mutagenic and/or carcinogenic substances make use of the chemical reactivity of the latter. Thus, compounds having electrophilic properties (e.g., alkylating and acylating agents) are treated with nucleophiles (e.g., water, bases, etc.), those being unsaturated can be reduced or oxidized, and so on.

Methods of destruction using a nucleophilic agent

Many mutagenic and/or carcinogenic compounds are endowed with powerful electrophilic properties (i.e., they have a high affinity for electrons) which explains in large measure their affinity for the nucleophilic sites of biological targets (e.g., proteins, nucleic acids, sulfur compounds, etc.).

Some of these compounds are very reactive and are hydrolyzed more or less rapidly when dissolved in water (water acts as a nucleophilic agent). This is the case for mustards (e.g., yperite), α-chloroethers, aziridines, and nitrosoureas (e.g., MNU, MNNG). Other compounds such as alkyl sulfates are hydrolyzed more slowly. In general, the hydrolysis of alkylating agents is accelerated in alkaline media (OH^-) or in the presence of various anions (e.g., SCN^-, S_2O_3, Cl^-, and especially $H_2PO_4^-$) (Table 2.5).

Destruction by contact with dilute sodium hydroxide solution. To a large excess of an ice-cold 10% aqueous solution of sodium hydroxide, carefully add the alkylating agent to be destroyed (an exothermic reaction may occur). Allow to stand overnight. Dilute with water and wash down the sink. If the compound is poorly soluble in water, it must first be dissolved in ethanol. Sodium hydroxide may be replaced by potassium hydroxide or even by ammonium hydroxide.

Compounds	Stability in water		Comments
	Hydrolysis	**T 1/2***	
Sulfur mustards (yperite) Nitrogen mustards (NH_2) or caryolysine	rapid rapid	5 min (37°C) 38 sec (20°C)	
Chloromethyl methyl ether (CMME) Bis (chloromethyl) ether (BCME)	very rapid rapid	1 sec (20°C)	Accelerated by OH⁻
Dimethyl sulfate Methyl methanesulfonate (MMS) Ethyl methanesulfonate (EMS)	slow slow slow	4.5 h (37°C) 5-10 h 10-15 h	Accelerated by : OH⁻ $S_2O_3^{--}$ $H_2PO_4^-$
Ethylene oxide Propylene oxide	very slow stable		accelerated by Cl⁻
β-Propiolactone	slow	3 h (25°C)	
Aziridine 2-Methylaziridine	rapid stable		Accelerated by : $H_2PO_4^-$ $S_2O_3^{--}$ H^+
Dimethylnitrosoamine (DMN)	fairly stable		10 % hydro- lysis after 14 days
N-Methyl-N-nitrosourea (MNU)	rapid	25 min	Accelerated by OH⁻
N-Methyl-N'-nitro-N- nitrosoguanidine (MNNG)	slow	35 h (37°C)	Accelerated by : $S_2O_3^{--}$ $H_2PO_4^-$ Cu^{++}

*Time required for disappearance of 50% of the substance.

Table 2.5   Stability of various carcinogenic alkylating agents in water.

This technique is effective for the destruction of:

–mustards
–haloethers
–sulfates and sulfonates
–nitrosoureas, nitrosocarbamates, nitrosoguanidines. The decomposition of nitrosoureas, nitrosocarbamates, and nitrosoguanidines in alkaline media is not recommended since this favors the formation of diazomethane, a highly volatile gas which has been found to be carcinogenic in rodents.

Alkylating agents such as alkyl sulfates, which are only slightly soluble in water, may also be destroyed by contact (1 hour) with a solution of 0.2 N sodium hydroxide in methanol. Thus, dimethyl sulfate may be destroyed by slow addition, with vigorous stirring, to a 0.2 N solution of sodium hydroxide or potassium hydroxide in methanol. After standing for 1 hour, the solution may then be stored for disposal.

$$(CH_3)_2SO_4 + 2\,NaOH \longrightarrow Na_2SO_4 + 2\,CH_3OH$$

Dimethyl sulfate

Destruction of α-haloethers.   For the hydrolysis of α-halomethylethers, in which formaldehyde and a hydroacid are released, it is preferable to use ammonia as the alkaline agent since it transforms formaldehyde into hexamethylenetetramine.

Under a properly functioning fume-hood, add 50 mg of haloether (e.g., chloromethyl methyl ether, bis(chloromethyl) ether, etc.) and 1 ml of alcohol to a round-bottom flask equipped with a magnetic stirrer. Add a 6% aqueous solution of ammonia dropwise to this homogeneous solution. Allow to stir at least 3 hours. Dilute with an excess of water and then store for proper waste disposal.

Destruction by contact with an aqueous solution of sodium hyposulfite. Many sulfur derivatives (e.g., hyposulfites, thiocyanates, mercaptans, etc.) may be employed to destroy electrophilic agents, though sodium hyposulfite appears to give the best results. Sodium hyposulfite is particularly effective for the destruction of sulfates, sulfonates, aziridines, and nitrosoureas.

Destruction of alkyl methanesulfonates.   To a large excess of a 10% aqueous solution of sodium hyposulfite ($Na_2SO_3 \cdot 10\,H_2O$), carefully add, while stirring manually, the alkyl methanesulfonate. At room temperature, the destruction of methyl methanesulfonate is complete within 1 hour, but approximately 20 hours is required in the case of ethyl derivatives. After destruction is complete, store for disposal.

Destruction of N-methyl-N′-nitro-N-nitrosoguanidine (MNNG).   Add the MNNG to a large excess of a magnetically stirred solution of 2% sodium hyposulfite in phosphate buffer (pH 8–9). Allow to stir for 1 hour. Store for proper waste disposal.

Destruction of aziridine.   To a large excess of a 0.5 M acetate buffer (pH 5) containing 10% sodium hyposulfite, carefully add the aziridine with stirring. Allow to stand for 1 hour and store for proper waste disposal.

Destruction by contact with an alkaline solution of thioglycolic acid.   Thioglycolic acid (or mercaptoacetic acid, $HSCH_2CO_2H$) is also effective in destroying most alkylating agents (sulfates, sulfonates, aziridines, etc.) by the formation of water-soluble salts. Strongly irritating to skin and mucosa, especially eyes, thioglycolic acid also possesses a disagreeable odor. It should thus be handled carefully under a properly functioning fume-hood.

Other methods of destroying carcinogenic substances

The IARC of Lyon, with the help of the Division of Safety of the National Institutes of Health (NIH) in the United States, has undertaken a systematic comparative study of the various methods of destroying carcinogenic substances. The techniques allowing destruction of eight categories of carcinogenic compounds have been published (after verification in an international control circuit):

Aflatoxins.   M. Castegnaro, D. C. Hunt, E. B. Sansone, P. L. Schuller, M. G. Siriwardana, G. M. Telling, H. P. Van Egmond, and E. A. Walker, *Laboratory decontamination and destruction of aflatoxins B_1, B_2, G_1, G_2 in laboratory wastes.* IARC, Scientific Publication No. 37. International Agency for Research on Cancer, Lyon (1980).

Nitrosoamines.   M. Castegnaro, G. Eisenbrand, G. Ellen, L. Keefer, D. Klein, E. B. Sansone, D. Spincer, G. Telling, and K. Webb, eds. *Laboratory decontamination and destruction of carcinogens in laboratory wastes: some N-nitrosoamines.* IARC, Scientific Publication No. 43. International Agency for Research on Cancer, Lyon (1982).

Polyaromatic hydrocarbons.   M. Castegnaro, G. Grimmer, O. Hutzinger, W. Karcher, H. Kunte, M. Lafontaine, E. B. Sansone, G. Telling, and S. P. Tucker, eds. *Laboratory decontamination and destruction of carcinogens in laboratory wastes: some polycyclic aromatic hydrocarbons.* IARC, Scientific Publication No. 49. International Agency for Research on Cancer, Lyon (1983).

Hydrazines.   M. Castegnaro, G. Ellen, M. Lafontaine, H. C. Van der Plas, E. B. Sansone, and S. P. Tucker. *Laboratory decontamination and destruction of carcinogens in laboratory wastes: some hydrazines.* IARC, Scientific Publication No. 54. International Agency for Research on Cancer, Lyon (1983).

Nitrosoamides.   M. Castegnaro, M. Bernard, L. W. Van Broekhoven, D. Fine, R. Massey, E. B. Sansone, P. L. R. Smith, B. Spiegelhalder, A. Stacchini, G. Telling, and J. J. Vallon. *Laboratory decontamination and destruction of carcinogens in laboratory wastes: some N-nitrosoamides.* IARC, Scientific Publication No. 55. International Agency for Research on Cancer, Lyon (1984).

Haloethers.   M. Castegnaro, M. Alvarez, M. Iovu, E. B. Sansone, G. M. Telling, and D. T. Williams. *Laboratory decontamination and destruction of carcinogens in laboratory wastes: some haloethers.* IARC, Scientific Publication No. 61. International Agency for Research on Cancer, Lyon (1984).

Aromatic amines and 4-nitrobiphenyl.   M. Castegnaro, J. Barek, J. Dennis, G. Ellen, M. Klibanov, M. Lafontaine, R. Mitchum, P. Van Roosmalen, E. B. Sansone, L. A. Sternson and M. Vahl. *Laboratory decontamination and destruction of carcinogens in laboratory wastes: some aromatic amines and 4-nitrobiphenyl.* IARC, Scientific Publication No. 64. International Agency for Research on Cancer, Lyon (1985).

Antitumor agents.   M. Castegnaro, J. Adams, M. A. Armour, J. Barek, J. Benvenuto, C. Confalonieri, U. Goff, S. Ludeman, D. Reed, E. B. Sansone, and G. Telling. *Laboratory decontamination and destruction of carcinogens in laboratory wastes: some antineoplastic agents.* IARC, Scientific Publication No. 73. International Agency for Research on Cancer, Lyon (1985).

These methods and others have been summarized in: G. Lunn and E. B. Sansone. *Destruction of hazardous chemicals in the laboratory.* John Wiley and Sons, New York (1990).

For these diverse families of compounds, the destruction methods tested generally rely on reduction (Ni-Al amalgam in alkaline media) and, especially, oxidation reactions among which may be cited:

- sulfochromic mixtures
- potassium permanganate
- hypochlorites (sodium, calcium, etc.)
- iodate, and so on

Methods such as ozonolysis, periodate oxidation, and oxidation of aromatic amines by sodium hypochlorite, have not been the object of verification and are thus not recommended.

Sulfochromic mixtures (potassium dichromate in sulfuric acid) easily destroy most organic compounds, but they are dangerous to handle (e.g., hexavalent chromium salts are allergenic, mutagenic, and carcinogenic) and their elimination poses a difficult problem (e.g., chromium salts are major pollutants of the aquatic environment). For these reasons, this technique, despite its effectiveness, has not been retained by the IARC.

Sodium hypochlorite (household bleach) reacts easily with aflatoxins (B_1, B_2, G_1 and G_2), hydrazines,* and aromatic amines. Polyaromatic hydrocarbons, however, are attacked very slowly (epoxide formation) and nitrosamines are totally inert.

---

*With hydrazine, the addition of a stoichiometric quantity of sodium hypochlorite leads to the formation of mutagenic substances. With dialkyl and diaryl hydrazines, sodium hypochlorite produces mixtures that contain nitrosamines. Hypochlorites [NaOCl and $Ca(OCl)_2$] are therefore not recommended for the destruction of hydrazines.

In the case of aflatoxins, a dichlorinated derivative is formed by addition to the furan double bond (in the 2,3 position). This derivative, which has mutagenic properties, must then be deactivated in a second step using acetone (with formation of a nonmutagenic dihydroxy derivative).

Particular techniques give good results with certain types of compounds:

Aflatoxins

–hot ammonium hydroxide (laboratory animal food and litter)
–lime (animal carcasses)

Polyaromatic hydrocarbons

–concentrated sulfuric acid after dissolving in DMSO

Nitrosoamines

–denitrosation with 3% hydrobromic acid in acetic acid
–formation of an oxonium salt by reaction with triethyloxonium tetra-fluoroborate followed by decomposition in alkaline media

Nitrosoamides

–denitrosation with hydrochloric acid in the presence of sulfamic acid
–denitrosation with hydrochloric acid in the presence of iron
–denitrosation with 3% hydrobromic acid in acetic acid (with elimination of volatile nitrosyl bromide)

Techniques for destruction with potassium permanganate in sulfuric acid

Oxidizing mixture ($0.3 M$ $KMnO_4$ + $3 M$ H_2SO_4). To 830 ml of ice water, carefully add, while stirring, 170 ml of concentrated sulfuric acid (d = 1.84). The reaction is exothermic. Cool the solution and slowly add, while stirring, 53 g of potassium permanganate. The purple solution is stored away from light.

Aflatoxins.   10 ml of the oxidizing mixture destroys 20 μg of aflatoxin in 3 hours at room temperature.

Polyaromatic hydrocarbons.   10 ml of the oxidizing solution destroys 5 mg of polyaromatic hydrocarbon (dissolved in 2 ml of acetone) in 1 hour at room temperature.

Nitrosoamines.   50 ml of the oxidizing solution destroys 300 μg of nitrosoamine in 8 hours at room temperature. It is preferable to allow the mixture to stand overnight.

Nitrosoamides.   10 ml of the oxidizing solution destroys 50 mg of nitroso-amide in 8 hours.

Hydrazines.   This technique may be applied to hydrazine, monomethyl-hydrazine, 1,2-dimethylhydrazine, and procarbazine. It cannot be used for the degradation of 1,1-dialkyl hydrazines and 1,1-diaryl hydrazines owing to the formation of large quantities of N-nitrosoamines. Moreover, this technique cannot be employed with DMSO solutions.

To 5 ml of the oxidizing solution, add 25 mg of the hydrazine derivative. Allow to stand overnight. Store for disposal.

Aromatic amines.   The aromatic amine is first dissolved:

- For benzidine and its derivatives, naphthylamines (1 and 2), and 2,4-dia-minotoluene: in HCl (0.1 mol/l) such that an amine concentration of 0.005 mol/l is obtained.
- For MOCA (3,3'-dichloro-4,4'-diaminodiphenylmethane): in H_2SO_4 (0.1 mol/l) such that an amine concentration of 0.001 mol/l is obtained.
- For 4-aminobiphenyl: in acetic acid such that an amine concentration of 0.001 mol/l is obtained.

The acidic amine solution (10 ml) is treated with 5 ml of potassium permanganate solution (0.2 mol/l) containing 5 ml of H_2SO_4 (2 mol/l). The mixture is left to stand 10 hours. Store for disposal.

General technique.   A solution of the carcinogenic compound in 3M H_2SO_4 or in an organic solvent (e.g., acetone, DMSO, DMF, etc.) is added to an excess of 0.3M potassium permanganate solution containing 3M H_2SO_4. The solution must remain purple, otherwise more 0.3M $KMnO_4$ solution must be added until the color persists. Allow to stand overnight. Store for disposal.

Remark.   Oxidation using a potassium permanganate–sulfuric acid mixture allows the destruction, without formation of mutagenic residues, of aflatoxins, a few nitrosamines, nitrosamides, polyaromatic hydrocarbons, and hydrazines as well as some antitumor agents (e.g., daunorubicin, methotrexate, vincristine, vinblastine, etc.).

2.31 The Particular Case of Destruction and Decontamination of Intercalating Agents of the Ethidium Bromide Family

Largely used in biochemistry for the detection of DNA, intercalating agents belonging to the family of phenanthridinium salts (e.g., ethidium bromide, propidium iodide, etc.) possess two free aromatic amine functional groups in their structures.

Ethidium bromide is a mutagenic substance whose carcinogenic activity has not as yet been demonstrated.

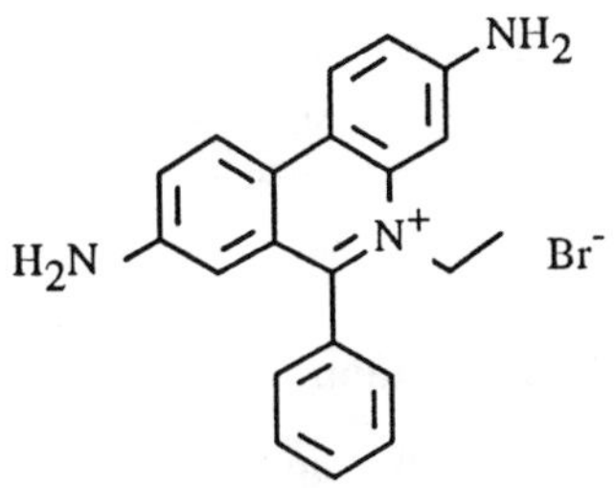

Ethidium bromide 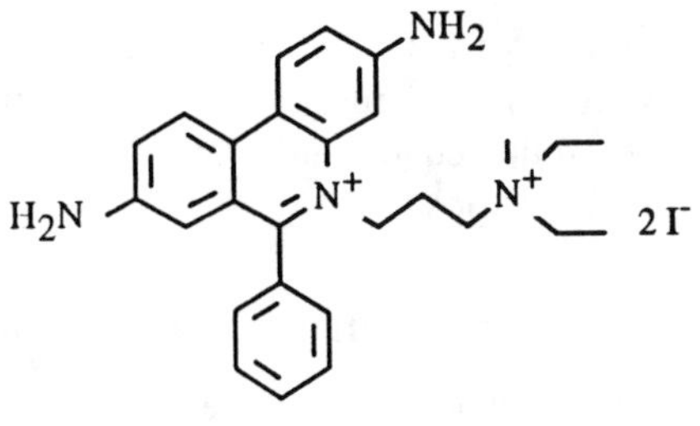   Propidium iodide

The destruction of ethidium bromide using the techniques of oxidation (i.e., sodium hypochlorite, potassium permanganate in acidic media) leads to the formation of residues possessing mutagenic activity, especially in the case of NaOCl.*

The best technique for the destruction of ethidium bromide (and probably, by analogy, propidium iodide) consists of deamination. The primary amine groups of this compound may be transformed into diazonium salts by the action of sodium nitrite. Reduction of the diazonium salts with hypophosphorous acid leads to formation of the corresponding hydrocarbon with release of nitrogen (Figure 2.1)

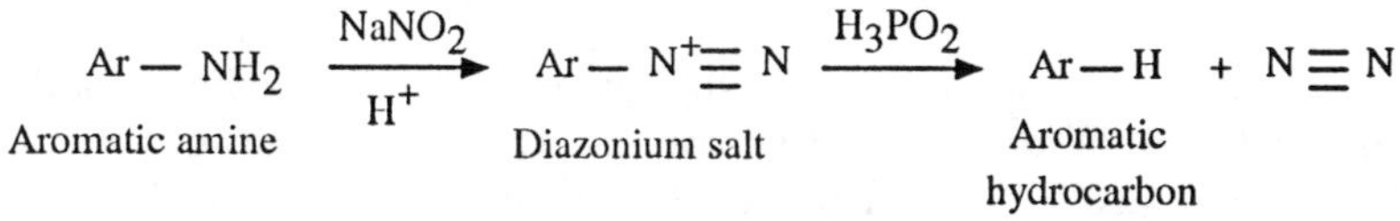

Aromatic amine   Diazonium salt   Aromatic
                                                hydrocarbon

Source: G. Lunn and E. B. Sansone. Ethidium bromide: destruction and decontamination of solutions. *Anal. Biochim.* 1987, **162**, 453–58.

Figure 2.1 Transformation of a primary aromatic amine into its corresponding hydrocarbon by reduction of the intermediate diazonium salt with hypophosphorous acid.

Destruction

This technique may be applied to aqueous solutions of ethidium bromide either in water alone, in TEB buffer, in Mops buffer, or containing cesium chloride.

To a 250 ml beaker equipped with a magnetic stirrer add

- 100 ml of an aqueous solution of 50 mg of ethidium bromide
- 20 ml of a freshly prepared 5% solution of hypophosphorous acid (stock solution: to 90 ml of water, add, while stirring, 10 ml of a commercial 50% solution of hypophosphorous acid), and 12 ml of a 0.5 N aqueous solution of sodium nitrite (34.5 g/l).

---

*After treatment with NaOCl, an ethidium bromide solution retains 20% of its mutagenic activity (P. Quillarde and M. Hofnung, *Trends Genet.* 1988, *4*(4), 89).

Allow to stir for several minutes. Gas is slowly evolved. Allow to stand for 20 hours. Add, while stirring, 1 g of urea in order to destroy excess sodium nitrite, followed by sufficient sodium bicarbonate to neutralize the solution. Store for disposal.

Methods for eliminating ethidium bromide from aqueous solution

Ethidium bromide may be eliminated from its aqueous solutions by two adsorption techniques using either blue-cotton or an Amberlite XAD-type ion exchange resin.

Adsorption on blue-cotton.   Blue-cotton is obtained by coupling cotton with a reactive dye, C.I. Reactive Blue 21 (copper trisulfophthalocyanine). One gram of blue-cotton contains 10 mg/mol of dye.
    To an aqueous solution of ethidium bromide, add an excess of blue-cotton. Adsorption is complete within a few minutes. Approximately 5 mg of blue-cotton per ml of solution adsorbs 1 μmol of ethidium bromide. After 15 minutes* of contact, the blue-cotton is removed from the solution, the excess liquid wrung out, and the cotton is destroyed together with other solid wastes.
    The adsorbed ethidium bromide may also be eluted from the blue-cotton using a solution of methanolic ammonia (50:1). Even when intercalated in DNA, ethidium bromide may be removed from aqueous solution by blue-cotton.
    A simplified technique, suitable for the treatment of large volumes of dilute ethidium bromide solutions (used to color agarose gels), consists of storing the solution to be destroyed in a carboy equipped with a stopcock at the bottom. The solution is then allowed to percolate slowly through a pad of blue-cotton placed in a funnel until the red color of the solution disappears. The saturated blue-cotton may be regenerated with a solution of ammonia in isopropanol (1:50). The concentrated solution of ethidium bromide can then be destroyed using the techniques described here.

Adsorption on Amberlite XAD.   To a 250 ml beaker equipped with a magnetic stirrer, add 100 ml of an aqueous solution (water, TEB buffer, Mops buffer, or cesium chloride solution) of ethidium bromide (100 g/ml). Add 2.9 g of Amberlite XAD-16 and allow to stir for 20 hours. Filter off the Amberlite and destroy together with other solid wastes.
    This technique, proposed by G. Lunn and E. Sansone (*Anal. Biochim.*, 1987) is, according to these authors, preferable to the blue-cotton method (higher adsorptive capacity, easier separation, and a much lower cost). According to B. L. Cohen (*Trends Genet.*, 1987, *3*(11), 308) this technique cannot be considered a routine procedure for the elimination of ethidium bromide solutions in biological laboratories.

---

*According to Lunn and Sansone (*Annal. Biochim.* 1987), contact should be maintained for 20 hours.

Adsorption on activated charcoal.   As with numerous aromatic and hetero-cyclic compounds, ethidium bromide, and apparently propidium iodide, are easily adsorbed on activated charcoal. In a technique proposed by O. Bensaude,* 1 mg of ethidium bromide in 100 ml of a saturated aqueous solution of CoCl is treated with 300 mg of activated charcoal. The mixture is intermittently shaken manually during 30 minutes at room temperature and then filtered on paper. The filter paper is combined with other solid wastes destined for incineration. The filtrate should no longer be red, otherwise the activated charcoal treatment must be repeated. The filtrate should also be checked for mutagenic activity.

This technique may also be used for the decontamination of glassware, workbenches, plastic trays, and so on. A variant of this method consists of percolating an aqueous solution of ethidium bromide through a disposable plastic column filled with activated charcoal. Once saturated, the column may be disposed of with other solid wastes.

Of all the methods described, that using activated charcoal seems to be the most suitable owing to both its simplicity and its low cost. This type of adsorption on activated charcoal may also be applied to azo dyes used in molecular biology (e.g., trypan blue, Evan's blue, etc.).

Remark

According to M. Fukunaga and L. W. Yielding,[†] propidium iodide is less toxic and, especially, much less mutagenic (after metabolic activation) than ethidium bromide. Since propidium iodide intercalates as easily as ethidium bromide with DNA, it could be tested as a replacement for this toxic substance.

---

*Trends Genet. 1988, 4(4), 89–90.
†Mutat. Res. 1983, 121, 89–94.

OTHER RISKS IN THE CHEMISTRY LABORATORY

CHAPTER

3

Biological Risks

While biochemists are frequently in contact with biological matter (normal or modified), it is no longer uncommon for chemists to engage in research at the interface of chemistry and biology and particularly of molecular biology. Because of this, in addition to risks associated with the use of increasingly sophisticated chemical substances (see Part 1), purely biological risks must now be contended with.

Biological risks in research laboratories first of all arise in the handling of wild-type, attenuated, or recombinant pathogens as well as in biological sampling, cell cultures, cellular manipulations (e.g., fusion, transfection), and animal experimentation.

3.1 Modes of Contamination

The major sources of contamination obviously come from the use, study, cultivating, mass production, or modification of pathogenic germs as well as sampling of contaminated biological material (e.g., organs, biopsies, blood and other biological fluids).

While contamination routes in research laboratories are relatively easy to envisage, the exact origin of the contamination is not always simple to determine. Among these routes of contamination may be cited:

- The cutaneous or percutaneous route: needle pricks during the course of injections (regardless of type or reason); cuts with sharp instruments (e.g., broken or chipped glassware, scalpels, etc.); bites or scratches by laboratory animals; projections or contact with damaged skin (e.g., scrapes, eczema, burns, etc.).
- The respiratory route: aerosols produced during the course of centrifuging, grinding, sonicating, homogenizing; manipulations with pipettes and syringes: working under inefficient or unsuitable fume-hoods.
- The conjunctive route: projection of contaminated material into eyes (particularly dangerous).

This chapter was prepared in collaboration with Drs. L. Mousel and G. Michaud (Pasteur Institute, Paris).

-The oral route: disregard for the basic rules of good laboratory practice (e.g., pipetting by mouth, eating or smoking in laboratories); touching the mouth with hands (e.g., nail biting, etc.).

In many cases, however, the exact mode of contamination is difficult to determine. In fact, in nearly 80% of the cases in which the sources of contamination were investigated, the origins could not be determined and were generally attributed to exposure to aerosols.

As a general rule, infectious aerosols produced by a variety of common techniques constitute a major risk in microbiological laboratories.

3.2 Classification of Wild-Type Pathogenic Agents

The classification of potential risks due to various pathogens is based on observed infections among hospital and laboratory personnel handling these agents. Among these numerous though similar classifications may be cited:

-United States: CDC-NIH; *Biosafety in microbiological and biomedical laboratories* (1969–1974, 1983, 1988).
-United Kingdom: *Code of practice for the prevention of infection in clinical laboratories and post-mortem rooms.* HMSO (1978).
-WHO: *Safety measures in microbiology* (1979).
-European Federation of Biotechnology, FEB report (1985).
-EEC: European community directive: *Protection of workers against risks associated with exposure to biological agents in the workplace.* 26 November 1990.
-In France, the AFNOR has adopted the FEB classification together with NFX 42-040 norms: *Microbial species commonly recognized as pathogenic in human,* (1990).

These classifications take into account both risks for the operator and risks for the community and the environment. These risks are a function of the pathogenicity and virulence of the microorganism, of its resistance, the modes of contamination, the existence of effective preventive and curative treatments, and the handling conditions.

The class 1 risks include agents which present no risks to those handling them or to the community. These are microorganisms which are most often nonpathogenic in human beings, for example, *Escherichia coli* (or *E. coli*) and *Bacillus subtilis*.

Class 2 includes microorganisms presenting a moderate risk to operators but limited, minor risks to the community. Viruses may be classed in this category. A preventive treatment always exists (e.g., vaccination). Some examples are: bacteria (*Bordetella pertussis, Borrelia, Clostridium tetani, Corynebacterium diphtheriae, Klebsiella pneumoniae, Legionella, Listeria mono-*

cytogenes, Staphylococcus aureus); virus (EBV, human herpes); fungus (*Aspergillus fumigatus, Candida albicans*); parasites (*Entamoeba histolytica, hymen-o-lepsis*, human and simian *plasmodium, Schistosoma, Trypanosoma brucei*).

Germs of class 3 expose handlers to serious risks and the community to moderate risks. The lesions due to infection are consequential. Prophylactic treatments sometimes exist. Some examples are: bacteria (*Bacillus anthracis, Brucella, Chlamydia psittaci, Francisella tularensis A, Mycobacterium tuberculosis*); virus (human hepatitis B, non-A non-B hepatitis, simian herpes B, simian smallpox, HTLV 1 and 2, HIV, yellow fever, dengue fever); fungus (*Blastomyces dermatiolis, Coccidoides immites, Histoplas*); parasites (*Echinococcus*, mammalian *Leishmania, Toxoplasma gondii, Trypanosoma cruzi*).

Class 4 agents present major risks to both handlers and the community. No treatments exist and risks due to propagation are high. Some examples are: bacteria: (none); virus (Lassa, Marrupo, Crimea, Congo hemorrhagic fevers, Marburg and Omsk viruses); fungus (none); parasites: (*Nagleria fowleri*).

These risk levels are only given as an indication.

In human beings, the risks of infection may vary according to the circumstances, but generally depend on three principal factors:

–the pathogenic potency of the infectious agent
–the severity of the contamination
–the state of the worker's immune system

The characteristics of the manipulated microorganism (e.g., bacteria, virus, etc.) are essential in evaluating risk and may be defined for a particular microorganism by:

–its virulence
–its pathogenicity
–its stability in biological media (e.g., intestine, etc.)
–its mode of transmission
–its endemic nature
–the possibility of an effective therapy (e.g., vaccines, antibiotics, chemotherapy, etc.).

3.3 Genetic Recombination

It is well-known that DNA contains the genes which constitute the basic units of heredity and the work program of cells. In order to effect genetic recombination, in which a foreign gene is introduced into a host cell thereby allowing the gene to be copied many times (i.e., amplification), it is necessary that the gene first be inserted into a vector. Many types of vectors are commonly used, for example, plasmids, phages, cosmids, viral vectors (DNA or RNA). Among the latter, the most frequently used are SV40, adenovirus, herpes-type virus vaccinia, and retroviruses.

The techniques of genetic recombination have, since 1976, been classed according to risks by the National Institutes of Health in the United States. This organization has specified the type of physical and biological containment appropriate for each type of risk. In general, these risks are a function of:

- the source of the DNA molecule
- the degree of purity of the DNA fragment
- the characteristics of the "host-vector" system
- the type of experiment to be performed
- the type of training received by personnel

While no accidents have yet been reported during the course of laboratory manipulations of recombinant agents, the infectious or other risks associated with genetically modified viruses have not been really studied in humans. Nevertheless, several recent studies in animals have led to many questions concerning the innocuity of such experiments. Particular attention should be paid to the risks which these may in fact present. Thus, in germ-free mice, especially those subjected to antibiotic treatment, persistent colonies of *E. coli* develop which are contaminated by the polyoma recombinant virus. These colonies are only transitory in normal mice.[*]

British scientists[†] have observed, after repeated application on lesioned mouse skin of a plasmid containing a DNA sequence originating from human cancer cells (T24 h ras oncogene), that blood vessel and lymphatic endothelial tumors appeared after several months. Cells extracted from the liquid contained in these tumors were shown to harbor the human ras gene.

Other researchers[†] have been able to induce tumor formation in chickens and mice by inoculating a chicken-derived viral oncogene. Moreover, the intramuscular inoculation into healthy, adult monkeys (*Macaca fascicularis*) of recombinant lambda bacteriophage DNA containing the proviral SIV Mac DNA sequence, led after several weeks to seroconversion of three of these animals[‡]. This seems to indicate that AIDS virus RNA may itself be infectious to those handling it, thereby suggesting that handling techniques should be reevaluated.

In summary, the potential dangers arising from recombinant DNA organisms appear to be in large measure the same as those of other microorganisms, that is, (1) the dangers of infection (e.g., bacterial, viral, etc.), (2) toxic effects associated with substances resulting from their metabolism (e.g., toxins, cytokines, etc.), (3) the dangers of environmental dissemination, but also (4) possible cancer promotion in tissue upon release of genes whose expression is directly or indirectly linked to the mechanisms of cell immortalization or transformation (i.e., oncogenes).

---

[*] C. Smith, E. Milewski, and M. M. Martin. The effect of colonizing mice with laboratory and wild type strains of *E. coli* containing tumor virus genomes. *DNA Tech. Bull.* 1985, *8*, 47-51.

[†] P. Braun. Naked DNA raises cancer fears for researchers. *New Scientist*, 6 October 1990.

[‡] N. L. Letwin, C. I. Lord, N. W. King, and M. S. Wyand. Risks of handling HIV, *Nature*, 1991, *349*, 573.

3.4 Animal Experiments

The handling of animals gives rise to two main sources of contamination.

3.4.1 Zoonosis

The potential risks associated with the use of various animal species are directly related to the endogenous pathogens that they carry. Some of these can be transmitted to humans as, for example, brucellosis.

The contamination of researchers or of personnel responsible for the care and upkeep of animals is a direct result of handling them (i.e., physical examination, assessment of mortality, weighing, various observations). Contamination may arise from air exhaled by these animals, from contact with excretions or other body fluids, or from aerosols produced during the handling of animals or their litter. As mentioned previously, bites and scratches can also result in contamination.

3.4.2 Inoculation

The second source of risk arises from the inoculation of infectious germs into animals. This risk is thus a function of the type of infectious agent utilized, although risk may also vary depending on the animal species.

3.5 Risks Arising from Cell Cultures

Regardless of whether one is dealing with primary cell cultures or with perfectly identified immortalized cell lines, the risks which these present may be of two types:

- Risks associated with the culture media and, in particular, with substances added to maintain or stimulate cell proliferation (e.g., growth factors, ionophores, cancer promoters, serums of various origins) or to avoid bacterial or fungal contaminations (e.g., antibiotics, antifungals);
- Risks associated with cell production itself; known or unknown, the substances produced may be:
  - various pathogens (particularly those involved in human pathologies) which are inherently present (and thus often unknown) or are introduced to produce immortalization (essentially viruses such as EBV, measles, CMV, adenovirus, HIV, hepatitis B, but also bacteria and parasites);
  - pharmaceutical substances, human proteins, recombinant vaccine proteins.

To these two major risk groups may be added:

- Risks related to the type of primary cells utilized (their origin, the modes

of sampling, and handling of explants destined for culture);
- Risks related to the products of expression of cloned genes: risk must be evaluated on the basis of the probability of penetration, of integration, of replication, and of expression in human cells.

Accidental contamination may occur at any step of an experiment but especially during the course of all manipulations which must be performed in the absence of strict containment (e.g., centrifuging, redistribution into fresh medium, sonication, animal injections, etc.).

3.6 Prevention: Biosafety Levels

The classification of pathogenic agents according to risk has led to the definition of biosafety levels (BSL) corresponding to each risk group (BSL1, BSL2, BSL3, BSL4). These biosafety levels have been the object of various international norms and regulations.* They are a function of:

- The type of laboratory, that is, its level of containment and its setup;
- The type of laboratory equipment appropriate for a given safety level (e.g., laminar flow hoods, type I, II, or III microbiological containment facilities);
- The definition and observance of good microbiological and virological techniques in general and of particular regulations specific to each safety level.

As for the prevention of chemical risks, the prevention of biological risks and the assurance of work quality depend on properly informing and training all personnel involved in this type of work. Hazard communication training, per OSHA, must be followed.

3.6.1 Biosafety Level 1 (BSL1)

This concerns a standard laboratory of $24\,m^3$ per person in which filtered air is renewed at the rate of $60\,m^3/hour$. The room should be easy to clean and decontaminate; workbenches should be smooth and resistant to acids, bases, and solvents. A basin should be reserved for washing hands and a cloakroom and rest area should be found nearby. An autoclave for the disinfection of equipment must also be available.

BSL1 does not necessitate any particular biological containment equipment for the protection of workers. Laminar flow-hoods are only used for the protection of cell cultures and their manipulation.

Good laboratory practice must be familar to all personnel and general

---

* *Biosafety in microbiological and biomedical laboratories.* CDC-NIH (1988).

regulations should be posted. Access to the laboratory is at the sole discretion of the unit head.

– Doorways into and within laboratories must be unobstructed to allow ease of circulation.
– Workbenches must be cleaned daily by the researchers and technicians using them.
– The treatment of wastes must be organized — incineration, autoclaving, or chemical decontamination (e.g., Javelle water, glutaraldehyde).
– A lab coat must be worn at all times within the laboratory.
– Pipetting by mouth is strictly prohibited. Mechanical pipettes must always be used.
– Drinking, eating, and smoking, as well as applying cosmetics, are to be prohibited. Food and drink must not be stored in the laboratory.
– Hands must be strictly kept clean, particularly before leaving the laboratory.
– A doctor must be advised of the work being conducted and the safety engineer should be informed of all incidents that occur.
– Insects and rodents should be regularly eliminated.

3.6.2 Biosafety Level 2 (BSL2)

Work involving class 2 pathogens (but rarely, and only under particular conditions, class 3 pathogens) must be conducted in BSL2 laboratories.

3.6.2.1 Laboratory Organization

The BSL2 laboratory may in fact be a closed BSL1 laboratory which may sometimes have an entry chamber. To avoid transporting contaminants toward the exterior, the laboratory should be under negative pressure. Alternatively, the entry chamber may be provided with a slightly higher pressure than the laboratory or access corridors. An autoclave must be made available on the same floor for the sterilization of biological waste. Futhermore, the international symbol for "biological risks" must be posted at the laboratory entrance.

3.6.2.2 Laboratory Equipment

The laboratory must be equipped with type I or II microbiological containment facilities. Under certain conditions, class 3-type germs may be handled, but in this case, an exterior ventilation conduit must be provided for.

3.6.2.3 Good Practice

In addition to the points just mentioned for BSL1 laboratories, BSL2 laboratories require that particular consideration be given to the protection of penetration routes in the body and to the risks related to aerosols.

–The type of microbiological containment facility to be used must be chosen as a function of the type of work to be effected. It must conform to local regulations concerning the protection of personnel, should be properly located, and regularly inspected and maintained.

–The wearing of designated lab coats, gloves, goggles, and face-masks, which are to be removed before leaving the laboratory, is essential. The working life of gloves is short and it should be remembered that they may become porous after cleaning with alcohol (PVC gloves) or produce allergies.

–Skin lesions must be protected and hands strictly kept clean

–Glassware, a frequent source of cuts, must be replaced by plasticware.

–Equipment containing infectious materials must be labeled with the international symbol for biological risks.

–Needles should be sheathed and eliminated in special, rigid containers.

–For injection or suction of infectious solutions, syringes having fixed needles or a needle lock must be employed.

–Centrifuge rotors must be equipped with an aerosol-free system.

–Workbenches must be decontaminated after each work session.

–Biological waste must be autoclaved.

–Specialized services should be called in for the disinfection of laboratories.

3.6.3 Biosafety Level 3 (BSL3)

The BSL3 laboratory is a strict containment facility allowing the handling of class 3 infectious agents which can cause serious, sometimes lethal, illness to workers and nonnegligible risk to the community.

3.6.3.1 Laboratory Organization

The laboratory should have a negative pressure of 6–10 mm and be leak-free. Entry should be through two entry chambers having a positive pressure with respect to the laboratory and external corridors. Air entering and leaving the laboratory is filtered through HEPA filters and is not recycled. Used liquids and water must be collected in containers and decontaminated (e.g., autoclave, chemical treatment) before being disposed of. One of the entry chambers should be used as a locker room. A double-entry autoclave, the simultaneous opening of doors being impossible, must be installed.

3.6.3.2 Laboratory Equipment

The laboratory should be equipped with type I, II, or III microbiological containment facilities. The latter must conform to regulations. All the equipment necessary for experiments must be found in the laboratory in order to avoid leaving before the end of the procedure.

3.6.3.3 Good Practice

The points just mentioned for BSL2 laboratories also apply to BSL3 laboratories.

- −Personnel having to work in BSL3 laboratories must receive specialized training.
- −The international symbol for biological risks as well as names of persons allowed access and those responsible for the work to be conducted must be posted at the entrance.
- −Access to the laboratory must be tightly controlled and the number of persons admitted must be limited.
- −The cleaning of the laboratory must be effected by the researchers and technicians working there.
- −All interventions in the laboratory by external services must be preceded by disinfection of the laboratory and of the microbiological containment facilities particularly before changing filters.
- −Special lab coats, overshoes, gloves, masks, and protective goggles must be worn.
- −No equipment must be taken out of the BSL3 laboratory before having been decontaminated (e.g., autoclaving, placing in hermetically closed containers before incineration, formol treatment of important equipment before repairing it or moving it to another laboratory).
- −Biological material which must be conserved outside the BSL3 laboratory should be placed in a closed container, disinfected before removal, and labeled with the biological risk symbol.

3.6.4 Biosafety Level 4 (BSL4)

BSL4 is required for all work involving the use of pathogenic agents which can menace the life of the experimenter as well as present significant risk to the community in case of dissemination.

3.6.4.1 Laboratory Organization

The BSL4 laboratory is a high-security containment facility, isolated from other laboratories, leak-proof, and under negative pressure. Access is strictly limited and controlled. Air entering and leaving the laboratory must be filtered twice. Windows should be sealed and unbreakable.

The laboratory should incorporate two entry chambers, separated by a shower. The first serves as a locker room while in the second, personnel dress in their work clothes. The process is inversed when leaving the laboratory. A porthole allowing disinfection (e.g., formol) of equipment leaving the laboratory must be provided for. As in the BSL3 laboratory, a double-entry autoclave must be installed for elimination of wastes. Furthermore, a tray allowing disinfection by immersion, fumigation, or spraying must be available for biological samples which are to be removed from the laboratory in sealed bags or containers.

3.6.4.2 Laboratory Equipment

All manipulations must be performed in a type III microbiological containment facility under negative pressure. Otherwise, a type II facility may be used if the experimenter wears an appropriate protective suit. Waste bags must be sealed thermoelectrically before being removed.

3.6.4.3 Good Practice

In addition to the points mentioned for BSL3 facilities, the following may be added:

- –Persons authorized to work in a BSL4 laboratory must have already received adequate training and must, in particular, have extensive experience in the handling of infectious materials.
- –Cleaning, changing filters, disinfecting, and eradication of insects and rodents must be performed by BSL4 personnel.
- –Upon entering the second entry chamber, the experimenter must don his specific and complete working attire (i.e., underwear, trousers, shirt, shoes, cap, mask, or protective suit).
- –No material or waste is to be removed from the BSL4 laboratory before having been disinfected. Live samples destined for conservation must be placed in a double, hermetically sealed container before disinfection and labeled with the symbol for biological risks.

3.6.4.4 Genetic Recombination and Prevention

In vitro recombinant DNA technology (commonly referred to as *genetic manipulations*) was originally the object of a "regulatory approach" and of particular preventive measures owing to the great uncertainty concerning the possible risks involved. It has currently been accepted that these preventive measures must be based on the same principles as those formulated for other infectious risks.

Recombinant genetic technology consists of the in vitro laboratory fabrication of new genetic elements (i.e., recombinant DNA) followed by their transfer to "host" microorganisms.

The most commonly utilized microorganism is *Escherichia coli* (or *E. coli*), the most frequently used strain of which is K12 *E. coli*. This genetically well-defined strain is nonpathogenic and can neither survive long in the human intestine, nor colonize it. It should however be noted that, in mice, pretreatment with an antibiotic favors intestinal colonization by K12 *E. coli*. For particularly dangerous experiments, special weakened strains of *E. coli* are available.

Three rules must be observed when performing recombinant DNA manipulations:

- −Choose the least pathogenic microorganism possible.
- −Use bacterial strains incapable of surviving in the intestine.
- −Employ plasmids as often as possible (these are circular, nonchromosomal DNA fragments of a bacteria that serve as vectors for the cloning of DNA in host cells and that cannot be transferred to other bacteria).

3.7 Conclusion

The handling of pathogenic agents and biological material generally presents risks to workers and co-workers as well as to the environment in case of dissemination.

Proper prevention of such risks depends on:

- −an awareness of these risks and, especially, of the precise moments when they appear
- −the training and informing of personnel
- −respect for good laboratory practice and safety regulations, particularly with regard to individual protection (i.e., wearing of lab coats, gloves, masks, safety goggles, etc.).

4

Laboratory Risks Associated with Nonionizing Radiation

4.1 Nonionizing Radiation

4.1.1 Electromagnetic Waves

An electromagnetic wave results from the association of a periodic sinusoidal electric field E and a sinusoidal magnetic field B of the same period which propagate in wave form. The principal characteristics of an electromagnetic wave are:

- its frequency γ in Hertz (Hz)
- its period $T = 1/\gamma$ in seconds
- its wavelength λ in meters in a vacuum:
  $\lambda = cT = c/\gamma$ (distance of propagation during one period)
  where c = speed of light
- its intensity I

4.1.2 The Classification of Electromagnetic Radiation

The spectrum of electromagnetic radiation extends from gamma radiation to long wavelength radiowaves. Depending on the effects of electromagnetic radiation on living matter, the following two classifications may be made:

1. *Ionizing radiation* (e.g., X-rays, etc.) whose energy is sufficient to ionize atoms and molecules.
2. *Nonionizing radiation* (e.g., UV, visible and IR radiation) whose weaker energy prohibits ionization.

Nonionizing radiation corresponds to electromagnetic radiation whose energy is too weak to produce ionization of biological molecules (e.g., water). This energy limit has been arbitrarily set at 13.6 eV, corresponding to a wavelength of less than 9.13×10^{-8} m (UV region). Among the types of

This chapter was prepared in collaboration with Drs. D. Folliot (EDF-GDF, Paris), J. C. Beloeil (ICSN, CNRS, Gif-sur-Yvette), and P. Maillard (Curie Institute, Orsay).

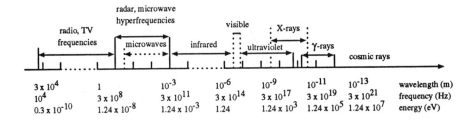

Figure 4.1   Electronic radiation spectrum.

nonionizing radiation may be cited:

–electromagnetic radiowaves (e.g., radio, television, microwaves)
–electromagnetic photowaves (e.g., IR, visible, UV)

Remark: In this distinction between ionizing and nonionizing radiation, particular mention should be made of laser radiation. The laser (*l*ight *a*mplification by *s*timulated *e*mission of *r*adiation) is a source of rigorously monochromatic and sometimes highly intense, coherent light, whose wavelengths vary from the UV to the IR as well as the visible. Their effects (essentially ocular lesions, burns, or fire risks) will not be discussed in this chapter.

4.1.3 Application of Electromagnetic Radiation in the Laboratory

While physical methods of analysis generally make use of numerous sources of ionizing radiation (e.g., X-rays) or nonionizing radiation (e.g., UV, IR, microwave, radiofrequencies), chemistry laboratories commonly employ UV sources of radiation (e.g., photochemistry). Chemistry and biology laboratories are, however, progressively using increasingly specialized spectral techniques:

–Ultraviolet (UV) and visible spectroscopy
–Infrared spectroscopy (IR)
–Nuclear magnetic resonance (NMR)
–Electron paramagnetic resonance (EPR)
–Circular dichroism (CD), and so on.

In addition, apparatus functioning with microwaves can now be found in laboratories, for instance, for chemical synthesis.
Only three areas will be considered here:

–Ultraviolet (UV) radiation,
–Magnetic fields (NMR, EPR),
–Radiofrequencies (microwaves)

While not actually radiation, ultrasonics will be briefly treated at the end of this chapter since their use in laboratories is becoming very common (e.g., ultrasonic cleaning baths, etc.).

4.2 Ultraviolet Radiation

4.2.1 Review of Some of the Physical Properties of Ultraviolet Radiation

Ultraviolet radiation is comprised of electromagnetic radiation having wavelengths between 4,000 and 40 Å. It extends from the violet end of the visible spectrum (4,000 Å) to the X-ray region with which it partially overlaps. For practical reasons, UV radiation may be subdivided into three regions:

  −near UV (A), from 4,000 to 3,000 Å
  −far UV (B), from 3,000 to 2,000 Å
  −extreme UV (C), from 2,000 to 40 Å

4.2.2 Risks of Exposure

In most laboratories, exposure to UV radiation may occur under the following circumstances:

4.2.2.1 *Microbiological Sterilization*

The bactericidal and germicidal effects of UV radiation begin at 2,500 Å (maximum at 2,600 Å) and UV is often used for the sterilization of the atmosphere and workbenches of microbiological laboratories. It should be noted that the efficacy of UV in the sterilization of the laboratory atmosphere is very controversial. It would in fact appear that these lamps are only effective over very short distances. Moreover, the wavelength of the emitted radiation varies with the age of the lamp, further reducing their bactericidal potency.

4.2.2.2 *Visualization of Chemical Compounds by Fluorescence*

UV lamps (hand-held or fixed) function on the principle of mercury vapor fluorescence. Among the resonance bands of mercury vapor radiation, those at 254 nm (short wave) and 366 nm (long wave) are commonly used for the detection of chemical compounds during the course of electrophoresis and chromatography (paper, column, and thin layer). Long wave (366 nm) UV lamps must be particularly guarded against.

4.2.2.3 Ultraviolet Spectroscopy

Ultraviolet spectroscopy is used to characterize and quantify chemical compounds having a UV-absorbing chromophore. Generally, the ligh source of spectrometers is well protected and the risk of exposure is usually the result of negligence (changing of lamps, etc.).

4.2.2.4 Photochemistry

Photochemistry, because of the power of the lamps and the often unprotected installations, presents the greatest risks of exposure.

4.2.2.5 Photocopiers

Many photocopiers use UV light sources and their proper protection should be verified.

4.2.3 Biological Effects

Depending on its wavelength and its intensity, UV radiation may have beneficial or deleterious biological actions on the human organism. These effects are due to the photochemical action of UV light.

The penetrating power of UV light is very weak; thus, for skin, corneal, and conjunctive tissue, it is only 0.1 mm. Overexposure to UV radiation provokes acute or long-term lesions.

> While IR radiation produces an immediate effect (thermal burn), UV radiation has a deferred action with a more or less long latency period.

The direct effects of UV light concern only the skin and eyes.

4.2.3.1 Effects on Skin

The direct, immediate effects on skin correspond to the classic "sunburn" and range from the simple cutaneous erythema (i.e., redness) which appears 6–12 hours after the initial exposure (attaining a maximum after 24 hours) to more serious symptoms (e.g., edema, etc.) with generalized disorders (e.g., fever, chills, nausea, delirium, prostration, etc.). An increase in pigmentation (i.e., tanning) may be observed later.

> The intensity of the erythema and of the eventual complications (i.e., edema, etc.) is proportional to the dose of UV light received.

Numerous photosensitizing chemical substances (e.g., furocoumarins, quinones, etc.) make the skin even more sensitive to UV radiation. Prolonged exposure of skin to UV light creates irreversible cutaneous lesions ultimately leading to "senile skin" (i.e., scaly skin with scattered pigmentation spots).

These lesions often precede precancerous lesions (i.e., actinic keratosis) or cutaneous cancers.

4.2.3.2 Effects on Eyes

It should be noted that UV (as well as IR) light is not detected by the human retina and the destructive effects it produces on the occular apparatus is rarely perceived immediately. This necessitates the use of strengthened precautionary measures.

> The absorption of UV radiation of wavelengths less than 310 nm by the exterior layers of the eye (i.e., cornea, conjunctive tissue) may provoke more or less serious eye damage.

Thus, conjunctivitis may appear 4–8 hours after exposure and last several days (1–5 days). While this inflammation is not in itself serious, a secondary infection must be watched for. Long-term exposure to UV light may lead to a partial loss of eyesight.

It is not evident that UV radiation can lead to eye cancers, but it should be remarked that blue-eyed persons are more susceptible to ocular melanomas. In conclusion, the biological effects of UV light depend more on the total quantity of energy absorbed (dose) than on the rate of its absorption.

> In case of an accident, a precise diagnosis can only be made if the dose of radiation received, as well as its physical characteristics, are known.

4.2.4 Protective Measures

Work areas where UV light sources are used (e.g., microbiology, photochemistry laboratories) must be well indicated by signs and equipped with efficient ventilation such that accumulation of ozone and nitrogen oxides is avoided. In sterile rooms, it is necessary to turn off the UV sources before entering.

4.2.4.1 Protection of Skin

Cutaneous protection is assured by the wearing of normal clothing (cotton, etc.).

4.2.4.2 Protection of Eyes

Ideal glasses, offering perfect protection and comfort under all conditions, do not exist.

> For effective protection against UV radiation, glasses having special, filtering glass and close-fitting lateral shields should be worn.

For protective filters, color should not be the only criterion of choice; in fact, glasses having an apparently identical color may absorb dangerous UV and IR radiation in a totally different manner. The wearing of inappropriate tinted glasses in the presence of harmful radiation is particularly dangerous due to the fact that, visible light being attenuated, the dilated pupil allows an even larger quantity of radiation to penetrate the retina. In all cases, organic (e.g., propionate, etc.), tinted glasses that absorb UV radiation should be worn. Certain glasses absorb UV, part of the visible as well as IR radiation. These glasses must be equipped with lateral shields adapted to facial contours in order to protect eyes against radiation coming from the side.

> Persons regularly exposed to nonionizing radiation (UV, IR, etc.) must regularly undergo eye examinations.

4.3 Magnetic Fields

4.3.1. Biological Effects of Magnetic Fields

As long ago as 1896, the biological effects of magnetic fields were investigated by Arsène d'Arsonval and, since this time, despite a very large number of studies, long-term effects are still largely unknown. While a static magnetic field does not appear to produce by itself detectable effects, the creation of an electric current in vivo due to the movement of an organism inside a magnetic field (as, e.g., in the case of NMR scanning in humans) may be the source of harmful effects on health.

The study of the long-term effects of magnetic fields has developed significantly in the past few years. It has been demonstrated that magnetic fields, even intense ones, are not mutagenic. Promoting effects are possible, however, which could explain the results of certain epidemiological studies which indicate that persons in contact with intense magnetic fields (e.g., cyclotron personnel) apparently have a slightly higher incidence of certain cancers (e.g., leukemia, central nervous system cancers, etc.). These results require confirmation, but would nevertheless suggest prudence.

4.3.2 Risks of Exposure and Prevention

The use of increasingly powerful NMR instruments, as, for example, for in vivo studies (i.e., Magnetic Resonance Imaging), requires that a few rules of safety be respected, both by those manipulating the instruments and those required to be in the vicinity of NMR equipment, especially the superconducting type.

4.3.2.1 Risks Associated with Magnetic Fields

> Never allow magnetic metallic objects (i.e., iron, nickel) to come within 3 m of a superconducting magnet (from 2.3–14 teslas).

> In particular, the transportation of cryogenic fluids in magnetic metal

containers (aluminum and nonmagnetic stainless steel pose no problem) must be avoided near superconducting magnetics, as is the use of metallic tools.

For persons wearing a pacemaker, it is forbidden to enter:

–the room that houses the magnet
–the rooms found above and below that housing the magnet if the latter has a field greater than 9 teslas
–such areas should be appropriately labeled to warn pacemaker wearers

Persons wearing metallic prosthetic devices (e.g., bolts or wires inside the body) should keep at least 3 m away from the instrument (danger of pain, etc.).

4.3.2.2 Other Risks

In addition to these effects associated with intense magnetic fields, three other risks must be taken into consideration:

–*Thermal risks*, which are secondary effects due to the emission of high intensity (100 watts to a kilowatt) electromagnetic waves (10–600 MHz) in the vicinity of the NMR spectrometer, particularly near the sample probe. These consist of thermal effects on living tissue by induction. Eyes are particularly sensitive to these thermal effects (cataracts, etc.).
–*Electrical risks*, associated with all equipment powered by electricity (220–380 V). If repairs must be made on the instrument, it should be remembered that very high voltages (several thousand volts) can be encountered inside.
–*Fire risks* which are possible in the presence of volatile, inflammable liquids (solvents such as carbon disulfide) due to the existence of microsparks in all electrical equipment.
–*Risks associated with the manipulation of cryogenic fluids* (e.g., liquid helium, nitrogen).

Electron paramagnetic resonance (EPR) utilizes hyperfrequency waves (microwaves) between 1 and 35 GHz (usually 9 GHz) of relatively weak power (0–250 milliwatts). The specific dangers arising from such weak microwaves appear to be fairly unimportant. The magnetic fields created around the source of emission are much less intense than those produced by NMR. All metallic objects (e.g., tools, etc.), however, should be kept at least 0.5 m from such instruments.

As with NMR, the risks associated with EPR are mostly related to:

–electrical problems
–fire
–risks arising from the use of cryogenic fluids

In particular, the use of photochemistry equipment (e.g., UV, visible and laser lamps) constitutes an additional source of danger encompassing:

–risks related to each type of irradiation

–electrical risks
–fire risks

4.4 The Use of Radiofrequencies

4.4.1 The Radiofrequency Domain

Within the domain of electromagnetic radiation, radiofrequencies comprise frequencies between 10 kHz and 300 GHz, corresponding to wavelengths between 30 km and 1 mm (see Figure 4.1). Radiofrequencies may be subdivided into two categories:

–Radiofrequencies per se: from 10 kHz to 300 MHz, used for radio, television, and so on.
–Hyperfrequencies: from 300 MHz to several hundred GHz. Microwaves are found in this category.

Wavelengths reserved for scientific, medical, industrial, and domestic use are found at $\lambda = 12.2$ cm ($\gamma = 2,450$ MHz).

4.4.2 Use in Chemistry

The use of microwaves in chemistry and especially in organic synthesis, is rapidly developing. Although household microwave ovens may be employed, this requires working with closed systems (Teflon-sealed tubes). The resulting elevated temperatures and pressures greatly accelerate chemical reactions, but these reactions are difficult to control and explosions can occur.

Special microwave units for chemistry laboratories are now commercialized and these may be used for:

–concentrating solutions
–warming liquid or pasty substances
–organic synthesis

With regard to organic synthesis, microwaves may be used to:

–generate reactive species (singlet oxygen, etc.)
–improve yields or purity of compounds obtained by various synthetic routes (nucleophilic substitutions, esterifications, hydrolyses, oxidations, rearrangements, organometallic syntheses, etc.).

The presence of solvents makes it difficult to control these reactions (explosions, fire, etc.). In order to avoid these situations, it is recommended to work without solvents (reactions on inorganic solid supports in dry media or phase transfer catalysis).

4.4.3 General Properties of Radiofrequencies

Many of the properties of radiofrequencies are similar to those of other electromagnetic radiation. Thus, the ratio between the wavelength and the dimensions of the structures with which the radiofrequencies interact determines the type of interaction. The specificity of radiofrequencies is therefore a function of their wavelength.

4.4.4 Biological Properties of Radiofrequencies

The effect of electromagnetic radiowaves on living matter is very complex and difficult to quantify. It is particularly difficult to establish simple relationships between the measurement of electromagnetic fields and the observed biological effects. For example, the quantity and distribution of absorbed electromagnetic energy is a function of:

- the frequency of the electromagnetic field
- the intensity of the field
- the configuration of the emitting source
- the composition and dimensions of the exposed tissues
- various environmental factors

4.4.4.1 Thermogenic Effects

Exposure to radiofrequencies first results in the heating of tissues. Above the thermal threshold, the effect produced corresponds to a local or general temperature increase that can exceed the possibilities of thermoregulation, leading to disorders. This temperature elevation is reversed by blood circulation which evacuates all or part of the calories produced. A poorly vascularized organ such as the eye lens is particularly sensitive to the thermogenic effect of radiofrequencies and the formation of a cataract is possible.

In the laboratory, care should be taken that microwave generators (microwave ovens, etc.) have no radiation leaks since, depending on the power of the source, superficial or deep burns may be produced.

4.4.4.2 Specific Effects

A current orientation of research in the field of electromagnetic radiation–living matter interactions concerns the effects of extremely low frequency (ELF) radiation situated between several hertz to 500 Hz, with 60 Hz frequencies having received the most attention. An international consensus has now been established in this area of bioelectromagnetism according to which weak nonthermal interactions exist which correspond:

In procaryotic organisms:

- To a diminution or an augmentation of cell growth (bacteria, yeast, etc.) as a function of frequency;

In eucaryotic organisms:

–To a perturbation in calcium flow, particularly at the level of cell membranes of the endoplasmic reticulum. This may lead to an acceleration of membrane peroxidation. Futhermore, magnetic fields modify calcium binding by calmodulin, an extremely important carrier protein for calcium in the cell;

–To a modulation in cardiac rhythm (production of arhythmia, etc.)

–To a modification of circadian rhythms, perhaps by an effect on the pineal gland leading to a decrease in melatonin levels.

Disorders caused in humans by radiofrequencies are of varied nature:

Effects on the neurovegetative system:

–physical asthenia (muscular disorders, etc.)

–mental asthenia (apathy, memory loss)

–headache

–sometimes tachycardia

These somewhat atypical disorders may also be provoked by other causes (stress, etc.).

Effects on the neuroendocrine system:

–Effects of microwaves on spermatogenesis have been described, but the observed disorders, including sterility, are reversible and may be due to thermal stress.

In general, it seems that clinical symptoms observed in humans are regressive and simply removing the subject from the field of action of electromagnetic radiation is sufficient to obtain progressive normalization of the affected functions.

4.4.5 Risks of Exposure and Prevention

The risks due to exposure to radiofrequencies, either domestic or professional, are of two natures:

–acute risks related to an accident

–long-term risks

4.4.5.1 Acute Risks

Few accidents have been reported up to now. One case describes the repair of a microwave oven which resulted in the irradiation of the repairman's face close to the source, resulting in ocular lesions and persistent headaches. It must be remembered that, when using a microwave oven, both electromagnetic fields necessary to heat matter as well as parasitic radiation are produced. Ovens should thus regularly be inspected for leaks.

4.4.5.2 Long-Term Risks

As mentioned in the section on magnetic fields, there is strong evidence that electromagnetic fields increase the risks of leukemia and of brain cancer. Epidemiological studies indicate that children are particularly sensitive to these effects. In the genesis of cancers, magnetic fields seem to act more as promoters, suggesting that repeated exposures over relatively long periods of time are necessary.

In summary, evaluation of the dangers to human health resulting from exposure to microwaves and radiowaves is difficult due to the extreme complexity of the relationships between exposure conditions and energy absorbed. Safety regulations vary from country to country depending on whether only thermal effects of radiation are considered or if specific effects unrelated to temperature are taken into account.

4.5 The Use of Ultrasonics

Ultrasonics, acoustic vibrations whose high frequencies (greater than 18,000 Hz) make them inaudible to the human ear (which reacts to frequencies between 20 and 18,000 Hz), are finding an increasing number of uses in chemistry and biology laboratories (sonochemistry).

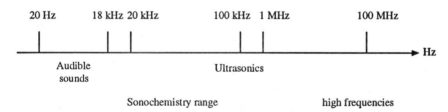

Ultrasonics have physical properties similar to those of sound: they propagate at the same speed in air (340 m/s at 20°C) and their speed increases with temperature.

4.5.1 Areas of Use

Ultrasonic waves are used in various chemical syntheses in order to accelerate certain reactions (e.g., polymerization, alkylation, etc.), but it is mainly their use in heating and in cleaning glassware (ultrasonic baths) that is becoming more widespread in both chemistry and biology laboratories.

4.5.2 Biological Effects

The biological effects of ultrasonic waves are due to thermal and mechanical effects as well as to cavitation (i.e., the appearance of bubbles in a sonicated liquid). These effects on cellular constituents are a function of the frequency and intensity of the ultrasonic waves. Only very high energy ultrasonics are capable of provoking detectable cellular alterations (e.g., chromosomal damage, etc.).

4.5.3 Preventive Measures

Ultrasonic baths should be placed in a protected, well-indicated area, for example, in a fume-hood if there is a possibility of formation of vapors or aerosols when a solution is sonicated.

Dirty glassware should be prewashed before being put in an ultrasonic bath. The latter should be turned off before glassware is added or removed and, as well, protective goggles and gloves should be worn. If the bath is in frequent or repeated use, earplugs should be worn. In order to diminish the noise, the ultrasonic bath should be covered and the condition of the cover periodically checked.

It should be noted that, in general, air is considered to provide good protection against ultrasonics though absorption of ultrasonics by air, like the absorption of high frequency noise, is fairly poor. Moreover, water constitutes an excellent means of propagating ultrasonics.

5

Laboratory Risks Associated with Radioelements

5.1 General Rules

The laboratory use of radioisotopes or of generators of ionizing radiation imposes the strict respect of particular rules.

5.1.1 Knowledge of Risks

The person competent in radioprotection must prove his knowledge of the possible risks (training certificate). This person must keep all personnel working routinely or temporarily in regulated areas informed of all dangers. The rules concerning the storage, handling, and disposal of radioactive material, as well as those intervening in case of accident, must be familiar to all personnel. To this end, persons regularly exposed to radiation should periodically renew and refresh their training.

5.1.2 Protective Measures

Both collective protection (e.g., workbenches, fume-hoods, glove compartments, rules, signs) and individual protection (e.g., gloves, lab coats, respirators, etc.) must be considered before radioelements are used in a laboratory.

This chapter was prepared in collaboration with Drs. J. C. Zerbib (CEA, Saclay, France) and G. Simonnet (INSTN, Saclay).

5.1.3 Monitoring

A person competent in radioprotection or, better, all persons handling radioelements, must monitor all radioactive sources entering and leaving the laboratory as well as eventual contamination (e.g., work areas, clothing, skin) by using a detector suited to the radiation emitted. The risks of irradiation due to contamination must be estimated as precisely as possible and proper verifications made as a consequence. Particular care should be taken to avoid irradiation of skin by beta emitters which deliver very high doses of radiation (40–80 mSv or 4–8 rems/hour for 1 μCi deposited on 1 cm^2 of skin).

5.1.4 Decontamination

Decontaminating material must be on hand and regularly verified. All those manipulating radioactive material must be aware of the techniques of primary decontamination.

5.1.5 Record Keeping

The person responsible for radioprotection must keep up-to-date records of all activities involving radioactive material (e.g., purchases, stocks, handling, wastes) and must ensure that wastes are removed in accordance with existing regulations.

5.2 Risks Associated with the use of Radioelements

The utilization of radioisotopes involves two major types of risks:

External irradiation, also referred to as "external exposure," which occurs each time someone is in the trajectory of radiation emitted by radioactive substances situated outside the organism.

Contamination, of which two forms are encountered:

– External contamination, corresponding to the deposition of a radioisotope on the skin; this results in irradiation of the extremely radiosensitive basal layer of the epidermis.
– Internal contamination, resulting from penetration of a radioisotope within the organism.

5.2.1 The Biological Effects of Irradiation

Ionizing radiation produces multiple cellular alterations among which mutagenic effects are the most dangerous.

A distinction must be made between nonrandom somatic effects, which are threshold effects produced at high doses — several grays (hundreds of rads) delivered over a short period — and nonthreshold random somatic effects.

Except for those affecting sex cells, mutations occurring in individuals are referred to as somatic. Depending on the dose received, the nature of the radiation, its intensity, and so on, the somatic effects of irradiation result in an increase in the number of cancers and a decrease in life span.

Genetic mutations affect the reproductive cells and are transmitted to the offspring of the person exposed to the radiation.

It must be remembered that the effects of irradiation are additive and cumulate during the entire lifetime.

Among the disorders resulting from the action of ionizing radiation may be cited skin ailments (e.g., radiodermatitis), cataracts, a decrease in fertility and an increase in cancers (e.g., leukemia, etc.).

Irradiation of a fetus in utero is particularly dangerous. Exposure of women of child bearing age should be limited as much as possible and special precautions should be taken for pregnant women in order to protect the fetus.

Human beings are naturally and continuously exposed to radiation and internal contamination by all types of ionizing rays. Natural radiation varies with geographical locations. Moreover, living in closed quarters exposes lungs to irradiation produced by natural radon released from building materials. This contributes the equivalent of 100 mrem of irradiation over the entire body.

In addition, the present conditions of civilization expose us to irradiation arising mainly from medical sources (diagnostic radiology, radiotherapy, etc.). Exposure to artificial radiation of medical origin averages about 0.7 mSv (70 mrem). It may be estimated that the total exposure of an adult to natural and artificial radiation averages an annual dose of 2.5 mSv (250 mrem).

The cumulative effects of weak doses are at present not well known. It is, however, acknowledged that all exposure to radiation, regardless of the level, carries a risk of cancer induction, either by a direct effect or by *a synergistic effect with a carcinogenic substance*. For example, it appears that ionizing radiation has a synergistic effect on the leukemia-promoting action of benzene.

5.2.2 Radiation Units

5.2.2.1 The Activity of a Radioactive Source

The official unit is the *becquerel* (Bq). This is the activity of a source having one radioactive transformation per second. The official unit previously used was the *curie* (Ci) corresponding to 3.7×10^{10} disintegrations per second (i.e., the approximate activity of a gram of radium).

$$1 \text{ Ci} = 3.7 \times 10^{10} \text{ Bq}$$

$$1 \text{ Bq} = 2.7 \times 10^{-11} \text{ Ci}$$

The microcurie (μCi) which is one-millionth of a curie, corresponds to 37,000 becquerels.

$$1 \mu\text{Ci} = 37 \times 10^3 \text{ Bq}$$

5.2.2.2 Units Measuring Effects on the Organism

The *Sv* (previously rem) is the unit which takes into account the fact that, for the same dose absorbed (i.e., the amount of energy delivered per unit of mass expressed in grays [Gy], previously in rads), all types of radiation do not produce the same biological effects. A quality factor, *QF*, is also introduced such that:

> QF = 1   for γ, X, β^- and β^+ radiation
> QF = 20   for α radiation
> QF = 10   for neutrons of unknown energy
>
> Sv = Gy × QF
> rem = rad × QF

The correspondence between new and old units is given by:

> 1 Sv = 100 rem
> 1 Gy = 100 rad

5.2.3 External Irradiation

All types of ionizing radiation (the term *radiation* is improperly used for all ionizing emissions, whether particulate or electromagnetic) are capable of provoking irradiation. This, however, depends on the nature and the energy of the radiation which determine its trajectory in air or in a given material. Thus, all α and β particles of energy less than 100 keV* penetrate air, and a fortiori, matter, so weakly that there is no risk of external irradiation. Consequently, it is indispensable to know, for each radioisotope used, the type of emission, the radiation energy, the thickness of materials necessary to stop or to attenuate the radiation.

It should, however, be noted that in case of internal contamination by these substances, the doses delivered may be large, especially in the case of α emitters. For X and γ emitters, diffusions and back-diffusions (reflections) may occur, resulting in irradiation.

Four parameters determine the dose absorbed for a given type of radioactive source:

–the activity of this source
–the distance from the source
–exposure time
–the nature and thickness of the screen

As with all dangerous manipulations, the exposure time to external radiation may be reduced by first conducting a "dry run."

---

* 1 keV = 1,000 eV. The electron volt is the energy unit used in nuclear fields. It corresponds to the kinetic energy of an electron accelerated by a difference in potential of 1 volt.

5.2.4 Internal Contamination

External contamination, that is, the presence of a radioactive substance on the skin or on mucous membranes, is an *essential risk* in chemistry and biochemistry laboratories.

Internal contamination may, however, occur by the following routes:

–the respiratory tract
–the digestive tract
–a direct route due to wounds (contaminated needle pricks)
–the transcutaneous route

The precautions in using nonsealed sources are the same as for all dangerous substances (vapors, liquids, powders, aerosols). Consequently, it is necessary:

–to isolate sources and contaminated objects
–to ensure a clean work space
–to use collective protective measures
–to use individual protective measures

It should be noted that the risks associated with the use of scintillation liquids* are at least as important as the risks of radioactive contamination.

Scintillation liquids are generally mixtures of a solvent and one or more scintillators (primary or secondary scintillators). The most frequently encountered solvents are toluene, xylenes (mixtures of three isomers), cumene (isopropylbenzene), pseudocumene (1,2,4-trimethylbenzene) and 1,4-dioxane. The scintillators are most often heterocyclic substances. Among primary scintillators, PPO (2,5-diphenyloxazole) is the most commonly used. Secondary scintillators are substituted diphenyloxazoles (POPOP and dimethyl-POPOP).

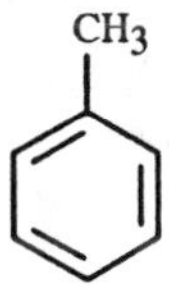

| Toluene | Xylene (mixture of 3 isomers) | Cumene | 1,2,4-Trimethylbenzene |

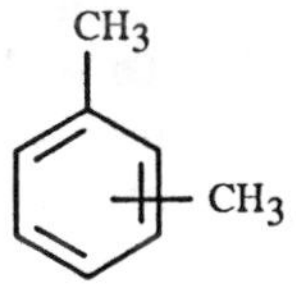

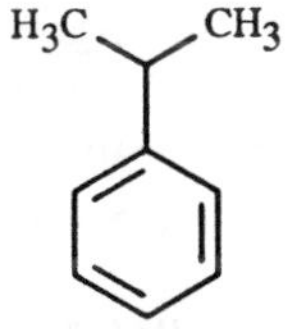

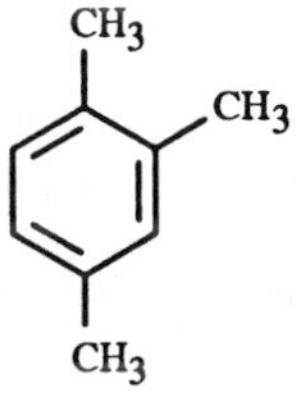

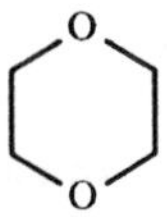

1,4-Dioxane

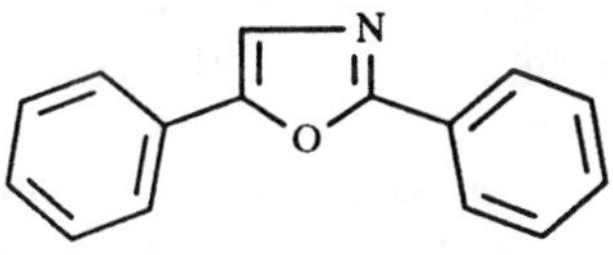

2,5-Diphenyloxazole (PPO)

*See Chapter 2.

1,4-Bis(5-phenyloxazol-2-yl)benzene (POPOP)

From a toxicological point of view, all aromatic solvents (e.g., toluene, xylene, cumene) are serious neurotoxins. Dioxane* irritates skin, the rhinopharynx, eyes, and the digestive tract (nausea, vomiting, etc.). Digestive problems may be complicated by hepatorenal ones (e.g., acute hemorrhagic nephritis, etc.). Experiments in rats show that, administered orally, dioxane produces nasal cavity and liver cancers. It is classified as a potential carcinogen in human beings.

The toxicity of scintillators (solutes) is not well known but appears to be quite low.

Ready-to-use scintillation liquids are now commercially available and these no longer contain toluene. Thus, Beckman's "Ready Safe" mixture utilizes pseudocumene and a polyarylalkane as solvents. These substances do not appear to present any major long-term toxicity and they are easily biodegraded.

For the various risks described earlier (radioactive substances, solvents, and scintillation products), frequent monitoring of contamination (e.g., work areas, skin, etc.) with appropriate detectors allows correct evaluation of risks and the extent of dissemination of activity.

The detector must be stored far from radioactive sources when not worn. It is useless in the case of α emitters, regardless of their energy, and of β emitters of energy less than 1 MeV.

If, following internal contamination, therapeutic measures are not rapidly taken, a certain amount of radioisotope enters the blood and becomes concentrated in one or several target organs. Thus, iodine concentrates mainly in the thyroid. For example, the inhalation of $1\,\mu$Ci of ^{32}P, a relatively low amount, delivers a dose of about $2\,$mSv to the target organs (bones), while inhalation of $1\,\mu$Ci of ^{125}I delivers a dose of $17\,$mSv to the thyroid.

5.2.5 Preventive Measures

All operations should be conducted as much as possible in trays lined with an absorbant covering (e.g., Benchkote paper) such that the dangers resulting from breakage or spillage are minimized. Liquid transfers should be done by automatic pipetting (pipettes with disposable tips or syringes, but never by mouth). For volatile substances, transfers may be conducted with an appropriate vacuum line equipped with a trap or filter. When working with particularly volatile products (e.g., ^{125}I used for the iodination of organic compounds such

---

* Dioxane must no longer be used in scintillation liquids because of its toxicity. It may be replaced by emulsifier-based scintillation liquids.

as peptides and proteins), all manipulations should be conducted in a glove-compartment.

Hands can be protected from contamination with gloves. Risks of contamination are significant if gloves are not properly removed. It is thus important to check for hand contamination after glove removal.

Activity must be measured at each step of the operation in order that the partition of activity between isolated products and wastes is known.

Perfectly adapted detection equipment must be available to allow frequent monitoring of work areas, equipment, and personnel (e.g., hands, shoes, clothing, hair, etc.).

As noted earlier, a dry run should be effected before radioactive elements are actually handled. (Although time factors do not play as important a role in determining contamination risks as they do in irradiation risks, a dry run allows observation of steps which may be particularly dangerous or which necessitate a modification of the procedure.)

The disposal of wastes is regulated. Special containers, either for liquids or for solids, must be on hand. Glassware must be decontaminated and rinsed before being reused. Special boxes must be used to dispose of contaminated syringes.

The rules of "nuclear" hygiene must be respected: eating, drinking, smoking, or applying cosmetics must be avoided in areas where nonsealed radioactive sources are used.

5.2.6 The Origin of Norms

The prevention of radioactive risks depends on the respect of certain norms. This system is based on the concept of an acceptable risk. The International Commission of Radiological Protection (ICRP) has established recommendations which are regularly revised as a function of progress in radioprotection. According to the ICRP, an acceptable dose of radiation is that which produces an acceptable risk both for the individual and the general population.

While the norms concerning external radiation have been expressed in a fairly simple form in keeping with physical characteristics, those relative to internal contamination by radioactive substances must be formulated for each of these substances considered individually.

5.3 Conclusion

The strict application of security measures concerning radioprotection may appear to be overly limiting. If it is remembered that the principal somatic risk of radiation is the appearance of mutations that may lead to the formation of cancers and that all doses of radiation increase this risk, then it is evident that

Radioprotection is not a useless bother in chemistry and biochemistry laboratories even when the activities or the radiotoxicities of the isotopes most frequently used are very weak.

APPENDIX

1

Principal Incompatible Products

Product Name	Formula	Incompatibility	Kind of Incompatible Reaction
Acetic acid	$CH_3{-}C{-}OH$ $\overset{\|}{O}$	CrO_3, $KMnO_4$, H_2O_2	Rapid oxidation
Acetone	$CH_3{-}C{-}CH_3$ $\overset{\|}{O}$	HNO_3, H_2SO_4 CrO_3	Rapid oxidation Rapid oxidation
Acetylene	$H{-}C{\equiv}C{-}H$	Ag^0, Hg^0, Cu^0, Mg^0 F_2, Cl_2, Br_2, I_2 O_2, O_3, $(NO)_x$, etc.	Explosive acetylides Rapid oxidation Rapid oxidation
Acrolein	$CH_2{=}CH{-}C{-}H$ $\overset{\|}{O}$	Strong acids Bases (NH_4OH, amines)	Violent polymerization
Alkali metals	Cs^0, Rb^0, K^0, Na^0, Li^0	Water	Exothermic formation alkaline hydroxide with hydrogen release
		Halogens (F_2, Cl_2, Br_2)	Formation of halide (explosive reaction)
		Alkyl halides (CCl_4, CH_2Cl_2, etc.)	Exothermic reaction
		Carbon dioxide (CO_2)	Combustion
		Sulfur (S_8)	Exothermic reaction
Ammonia	NH_3OH	Silver nitrate, silver oxide, etc.	Formation of explosive silver nitride (AgN_3)

Note: To distinguish their different states of oxidation, metals in their fundamental state are noted with a superscript zero, e.g., Mg^0.

Continued on next page

Product Name	Formula	Incompatibility	Kind of Incompatible Reaction
		Bromine	Formation of explosive nitrogen tribromide
		Alkyl sulfates (dimethyl-, diethyl-)	Extremely exothermic reaction
Bromine	Br_2	Unsaturated compounds (olefins, alcines, etc.)	Exothermic bromination
		Carbonylated compounds possessing a hydrogen in alpha (aldehydes, ketones)	Exothermic bromination
		Esters (diethyl oxides)	Combustion
		Metals (Al^0, Hg^0, Ti^0, etc.)	Exothermic formation of bromide
		Ammonia, ammonium hydroxide	Formation of explosive nitrogen tribromide
		Hydrides (SiH_4, PH_3)	Exothermic bromination and combustion
Chlorine	Cl_2	Organic materials (rubber, etc.)	Exothermic chlorination
		Diethyl oxide, tetrahydrofuran Dimethylformamide	Combustion
		Hydrazines	Formation of hydronitric acid
		Ammonia	Formation of explosive nitrogen trichloride
		Phosphorus Hydrides (AsH_3, PH_3, SiH_4, B_2H_6, etc.)	Exothermic chlorination
		Silicones	Exothermic reaction
Chromic acid	CrO_3	Flammable liquids (alcohols, ketones, acids, etc.)	Rapid oxidation
		DMF, pyridine, etc.	Violent reaction
		Sulfur (S_8)	Spontaneous combustion
Dimethyl-formamide (DMF)	CH$_3$ \ N—C—H / ‖ CH$_3$ O	Thionyl chloride	Exothermic reaction
		Chlorine	Exothermic reaction
		Carbon tetrachloride	Exothermic reaction
		Sodium hydride Sodium tetrahydroborate ($NaBH_4$)	
		$KMnO_4$, Br_2, Cl_2	Exothermic reaction (combustion)
Dimethyl-sulfoxide (DMSO)	CH_3—S^+—CH_3 | O^-	Acyl chlorides	Formation of formaldehyde (polymerization)
		$POCl_3$, PCl_3, SCl_2, SO_2Cl_2, $SOCl_2$, etc.	

Continued on next page

Product Name	Formula	Incompatibility	Kind of Incompatible Reaction
		Perchlorates (of magnesium, silver, mercury, chromium, etc.)	Solvation of the perchlorates by the DMSO
		Sodium hydride	Formation of dimethyl-sulfinylic anion (exothermic reaction)
		Solid $KMnO_4$	Combustion
Mercury	Hg^0	Acetylene	Formation of mercury acetylide (explosive)
		Ammonia	
		Br_2, Cl_2	Formation of halide
		Na^0, K^0, Li^0	Formation of amalgam (exothermic)
		Sulfur (S_8)	Exothermic reaction
Nitric acid	HNO_3	Organic combustible materials (cotton, wood, etc.)	Rapid oxidation (combustion)
		Alcohols (methanol, ethanol, ethylene glycol)	Formation of nitric esters (rapid oxidation)
		Ketones (acetone, methylisobutylketone, etc.)	Rapid oxidation
		Acetic anhydride	Formation of acetyl nitrate
		Aromatic amines (aniline, toluidines, etc.)	Rapid oxidation
		Hydrazines	Rapid oxidation
		Hydrides (PH_3, ASH_3, SH_2, SeH_2, etc.)	Rapid oxidation
Hydrogen peroxide	H_2O_2	Combustible organic materials (fats, etc.)	More or less rapid oxidation according to the concentration of H_2O_2
		Alcohols (methanol, ethanol, glycerol, etc.)	
		Acetone	Formation of explosive cyclic peroxides
		Carbocyclic acids (formic, acetic, tartaric acids, etc.)	Peracid formation
		Nitromethane	Explosive mixture
		Hydrazines	Rapid oxidation
		Metals (Ag^0, Cr^0, Co^0, Mn^0, Pb^0, Pa^0, Pt^0, etc.)	Decomposition
Perchloric acid	$HClO_4$	Organic combustible materials (wood, paper, cotton, etc.)	Formation of perchloric esters (rapid oxidation)

Continued on next page

Product Name	Formula	Incompatibility	Kind of Incompatible Reaction
		Dehydrating agents (acetic anhydride, P_2O_5, H_2SO_4, etc.)	Formation of perchloric anhydride (Cl_2O_7)
		Alcohols (methanol, ethanol, glycols, etc.)	Formation of perchloric esters (R—O—Cl_3)
		Sulfoxides (DMSO, dibenzylsulfoxide, etc.)	Formation of perchlorates
Phosphorus	P_4	Oxygen, air	Spontaneous combustion
		Oxidizing compounds ($KClO_3$, $KMnO_4$, etc.)	Explosive reaction
		$MgClO_4$	Explosive reaction
		F_2, Cl_2, Br_2	Combustion
		Alkaline hydroxides (KOH_3, NaOH, etc.)	Formation of spontaneously combustible phosphine
		Animal carbon, charcoal	Spontaneous combustion
Potassium permanganate	$KMnO_4$	H_2SO_4	Formation of dimanganese heptoxide (Mn_2O_7) (exothermic reaction)
		HCl	Exothermic reaction
		Acetic acid	(combustion)
		Acetic anhydride	Exothermic reaction
		Polyols (glycols, glycerol)	Combustion
		Aldehydes (formaldehyde, benzaldehyde)	Exothermic reaction (combustion)
		DMSO, DMF	Exothermic reaction
		Phosphorus, sulfur	Violent reaction
Primary and secondary amines	R—NH$_2$R—N—R' $\mid$ H	Hypochlorites (NaOCl, etc.)	Formation of chloramines
Sodium hypochlorite	NaOCl	Acids	Dichlore compound release
		Primary or secondary amines	Formation of unstable chloramines
		Alcohols (methanol, etc.)	Formation of unstable alkyl hypochlorite
		Ammonium salts (sulfate)	Formation of explosive nitrogen trichloride
Strong mineral bases	NaOH, KOH	Water	Exothermic dissolution
		Strong acids	Exothermic neutralization
		Trichloroethylene	Formation of self-combusting dichloroacetylene

Continued on next page

Product Name	Formula	Incompatibility	Kind of Incompatible Reaction
Strong mineral acids	HCl, H_2SO_4, HNO_3	$NaOH$, KOH, $HONH_4$	Exothermic neutralization
		$NaOCl$	Dichlore compound release
		$NaCN$, KCN	Release of toxic hydrocyanic acid
		NaN_3	Release of toxic hydrazoic acid
Sulfur	S_8	Alkali metals (K^0, Na^0, etc.)	Exothermic reaction
		Alkaline earth metals (Ca^0, Mg^0)	Exothermic reaction
		Mercury	Exothermic reaction
		Iron, copper, zinc	Exothermic reaction
		Tin in a divided state	Exothermic reaction
		CrO_3	Spontaneous combustion
Sulfuric acid	H_2SO_4	Water + fuming sulfuric acid	Violent reaction
		$KMnO_4$	Formation of $HMnO_4 + Mn_2O_7$
		$KClO_3$	Formation of ClO_2
		Polymerizable compounds (acrylonitrile, cyclopentadiene, etc.)	Explosive polymerization
		Nitrated compounds (nitromethane, nitrobenzene, etc.)	Exothermic reaction

2

Threshold Limit Values

The first experimental scientific works having to do with human toxicity to industrial products took place in the 1880s at the University of Würzburg in Germany, by Professor K. B. Lehmann.*

World War I gave a great deal of impetus to this type of experimental study. Two main branches can be distinguished:

1. The focus on chemical weapons (more than 4,000 products studied, 54 actually used),
2. The birth, in the United States, of a national chemical industry designed to compensate for the lack of German supplies.

The first list of *Maximum Allowable Concentrations* was published in the United States in 1937. It was in 1948 that an association of experts, the American Conference of Governmental Industrial Hygienists (ACGIH)—created in 1936—developed the idea of threshold limit values (TLVs). The publications of the ACGIH on the subject of threshold limits are still authoritative.

A2.1 Foreign Threshold Limit Values and the International Context

The best-known threshold limit values are those of the ACGIH. This privately held association publishes these values in a small booklet which is then

---

Jean-Claude Zerbib (CEA, Saclay).

furnished to its members. In 1988, the INRS published its limit values, which were repeated in the United States by the Occupational Safety and Health Administration (OSHA) in the form of regulatory limit values. In Germany, a commission for the study of substances dangerous to one's health in the workplace (MAK-Kommission) established a list of TLVs, repeated by the Labor Ministry of West Germany each year since 1969. In the former Soviet Union, the TLVs were published by the Ministry of Health in standards (in effect since 1 January 1977) that include additives.

A2.1.1 Different Limit Values

The diverse designations of limit values do not always overlap, and their levels are sometimes different.

A2.1.1.1 Denominations of the ACGIH

-TLV = threshold limit values
-TLV-TWA = TLV-time weighted average (averaged over 8 hours/day)
-TLV-STEL = TLV-short-term exposure limit (short-term exposure limit value is 15 minutes)
-TLV-C = TLV-ceiling (ceiling values that must never be exceeded, even for a short time)

A2.1.1.2 Denominations in Germany

MAK value = Maximale Arbeitsplatzkonzentration (maximum allowable concentration in the workplace averaged over 8 hours/day). For short-term exposures, the maximum concentration (brief peaks in concentration) are given in MAK numbers.

A2.2 Threshold Limit Values in French Regulations

The first average exposure values (AEV) set in regulations in France concerned the protection of workers against the risks presented by ionizing radiation. The decree of 15 March 1967 allowed for 257 radioactive products—in soluble or insoluble form, ingested by respiratory or digestive means—in AEVs calculated for a "work day" of a length of 8 hours. The basis of the calculation rests on a maximum ingestible quantity in 1 year of work (8 hours/day, 5 days/week, 50 weeks/year).

Regarding toxic chemicals, the first set AEV concerned benzene (decree of 11 November 1973) and the second asbestos (decree of 17 August 1978). This last was made more stringent in 1987 and then in 1992. Finally, in 1980 an AEV was set for monomeric vinyl chloride (decree of 12 March 1980). Note that the first products to be included in French regulations were carcinogenic substances.

A2.3 The Necessity of Setting Threshold Limit Values

Most workers breathe in these more or less toxic products in the course of their professional duties.

An important percentage of workers in the "base chemical, artificial, and synthetic fibers" sector (69%) describe inhaling of toxins. More generally, 48% of all workers, whatever their sector of activity, describe inhaling of powder, according to an inquiry of the Ministry of Labor (1985).

In addition, if one considers the annual statistics on work-related illnesses of the Caisse Nationale d'Assurance Maladies des Travailleurs Salariés (CNAMTS), the French National Salaried Workers Insurance Fund, representing nearly 4,200 annually "recognized" work-related illnesses, it is evident that the inhalation of different powders is responsible for more than one of three officially recognized work-related illnesses.

In addition, the publication of TLVs specifically concerning carcinogenic substances was initiated with a circular on 14 May 1985, that exposed the various elements of the Labor Ministry's adopted position on this issue. Three other circulars relating to carcinogenic substances were also published (circulars of 12 May 1986; 20 November 1986; and 14 March 1988). These circulars indicated not only the products, agents, or procedures where carcinogenity had been proven in humans, but also those where, in the absence of conclusive epidemiology, there exist adequate proofs of carcinogenicity for animals.

It is therefore necessary to define a standard of air quality for respiration in work areas.

A2.4 Circulars on "Dangerous Substances" and the Regulations

It was in 1981, within a Specialized Commission of the Conseil Supérieur de la Prévention des Risques Professionnels (CSPRP), the French Supreme Council on the Prevention of Work-Related Risks, that two work groups were created. One was devoted to the study of the available data on noncarcinogenic dangerous substances, and the other analyzed the available data on carcinogenic substances and products. Once examined by the special commission, the proposed TLVs were published progressively in the form of circulars.

Two types of admissible values were considered: (1) TLVs, which are the values measured for not more than 15 minutes, and (2) ATVs, determined for the length of a work day (8 hours).

The first circular relating to noncarcinogenic dangerous substances was published on 19 July 1982 (37 substances). It was followed by six other circulars (6 March 1983; 11 December 1983; 10 May 1984; 5 March 1985; 5 May 1986; and 13 May 1987).

With all of these circulars TLVs are available for 505 substances. These values are largely based on the values of the American ACGIH. Let us also note that the "envelope" of regulatory values was set by the decree of 7 December 1984, which introduced regulatory limit values for powders, gases, and aerosols to the Labor Code (Article 232-1-5).

> In specific polluted locales, the average concentrations of total and alveolar powders in the atmosphere inhaled by one person, evaluated over a period of 8 hours, must not pass, respectively, 10 and 5 milligrams per cubic meter of air. Particular measures taken in application of the second paragraph of Article L231-2 should the need arise for:
>
> > –other limits than those set in the first paragraph for certain kinds of powders;
> > –limit values for substances such as gas, aerosols, liquids, or vapors, and for climatic parameters.

From a formal point of view, only the limit values set by decree carry strong judicial weight; the TLVs and ATVs published by circular are only guidelines. However, the labor inspector is empowered to give "formal notice" in case of characteristic overdose following the TLVs and ATVs under Article L231-5 of the Labor Code.

A2.5 Measurement of Limit Values

Setting limit values, guiding or regulatory, poses the problem of controls for the atmosphere in the workplace.

When the Labor Ministry defines, by way of decree, the provisions relating to a physical agent or dangerous chemical (noise, ionizing radiation, benzene, asbestos, vinyl chloride, lead, carbon monoxide), it also provides a laboratory verification procedure. Each year, it grants a license to the laboratories according to the opinions of a special commission of the CSPRP, for agreements for a period of 1 year (new laboratories) or 3 years (renewal).

In a great number of cases, the INRS plays an important role in the distribution of licenses (evaluation of the quality of the files, or results of quality tests) delivered to the laboratories. The test results furnished by the laboratories are not free. Depending on the case, the prices are relative to the collection and measurement of a sample, or are fixed by half-day or day or testing.

As for other toxic products that have not been the object of specific decrees, the INRS has validated, for its own usage and that of the Services de Prévention des Caisses Régionales d'Assurance Maladie (CRAM), the Prevention Service of the Regional Health Insurance Fund, some methods of sampling and analysis.

It is possible to contact the INRS or the more than able services of the CRAM, who possess an "interregional chemical laboratory" if you wish to take samples with a view to measuring the concentration of a toxic product in the air. These tests are free.

In certain cases, you can conduct an individual survey of the thresholds of one or several toxic agents. The sampling takes place in the middle of a static system (gas-badge type) or a dynamic system. The latter consists of a small portable pump attached to a sampling tube (toxic gas) or to a filter (powders, aerosols).

The dosage of toxic substances thus sampled vary in different ways in the laboratory. The individual dynamic samples lead to more reliable evaluations of the incorporated charge than those deduced from the measurements of the "gas-badges."

A2.5.1 The Biological Exposure Survey

The introduction of toxic products into an organism is not made exclusively by inhalation. It may also occur by ingestion, through the pores of the skin, or even by direct contact (passage of the product through a scratch by a tainted object), or through an absorbant membrane.

The evaluation of the total quantity of the product taken in consists, therefore, in the measurement of the toxic substance itself, or of one of its biotransformation products (metabolites) in several different biological assays (blood, urine, feces).

One thus measures the *indicators of internal exposure*, also called *biological markers of exposure*. Following the period of biological activity of the substance, the result of the tests will yield information on the quantity of the recently ingested product or on the internal dose of the organism, after a long-term internal contamination.

The biological exposure survey constitutes a relatively recent preventative tool as far as toxic chemicals are concerned. For several decades, however, controlling radioactive contamination of the organism has been enacted, notably, by the CEA and EDF. The biological exposure survey for toxic chemical products made its entry into French law with the decree of 1 February 1988 on lead.

3

The European Framework Relative to the Prevention of Biological and Toxic Risks in the Workplace

A3.1 Summary of the Institutional Framework

Three treaties constitute the institutional framework of the European Community.

– The CECA treaty (European Community of Carbon and Steel), adopted in 1951 which, as the name indicates, was designed to promote the economic expansion and recovery of the carbon and steel industries after World War II.
– The CEE treaty (EEC or European Economic Community), signed in 1957, is the most important of these treaties; its goal is assuring the economic development of Europe by the establishment of a common market. This treaty was modified in a substantial way by the Maastricht Accords of February 1992, whose aim is the establishment of the European Community (EC).
– The EURATOM treaty (treaty instituting the European Community of Atomic Energy), also adopted in 1957, has the mission of, in the terms of Articles 1 and 2, assuring a "rapid" growth in the nuclear industries by the establishment of common standards in the area of security, by the development of technological research, and by promoting the spread of knowledge.

To attain these goals of strong economic growth, maintaining "community" or "harmonized" policies is necessary.

Jean-Luc Pasquier. Subcommittee on Work Conditions and Protection Against the Risks of Work, Ministry of Labor, Employment, and Professional Education, Paris.

It is thus, in being governed by the EC treaty, that these common policies must be implemented — and, moreover, have been — particularly in the areas of customs tariffs, agriculture, and transportation; and that harmonized policies must be developed concerning the elimination of impediments to free trade.

Depending on the case, the EC is empowered to issue regulations or directives — the directive being, hypothetically, the legal instrument of the harmonized policy. It generally has the aim of defining a precedent, sometimes quite precisely, while leaving to the member states the job of amending their own legislation or regulations, and of choosing the most appropriate methods of enaction.

Several institutions may step in, all with equal jurisdiction:

–The European Community Council, composed of heads of state or government, and ministers, who guarantee the representation of their states and who possess the largest share of the legislative power.

–The European Community Commission, which is the guarantor of treaties' enactment, and which has the power to make proposals for the EC. It is an entirely separate institution, that plays a much more important role than that of national administration.

–The European Parliament, issuing directly from universal suffrage, votes on the budget and plays, since the Unification Act and in what is called a position "of cooperation," a joint role with the council in drawing up directives.

–The Court of Justice at Luxemburg, which is the repository of EC law.

A3.2 Community Policy on Chemical and Biological Agents Present in the Workplace

For nearly 20 years, but particularly in the past decade, French legislation and regulation in the matter of prevention of chemical, and more recently, biological risks has essentially originated in texts taken from the outlines laid out by the EEC: the European directives.

The recent adoption of the European Unification Act and the integration of the common market planned for 1993 reinforces this tendency, by facilitating the drawing up of directives, notably by the mechanism of adoption by a qualified majority. All the directives on this subject are based on the treaty instituting the European Community (Articles 100, 100A, and 118A).

The initial objectives of the EEC treaty were of an economic nature; as a result, the policies developed in the social domain, especially in the areas of work safety and hygiene, were originally designed to encourage trade. In effect, this abolishes many unjustified nontariff barriers and allows the meshing of national legislations which, explicitly or implicitly, could raise obstacles to free trade, or which could lead to distortions in areas of agreement.

For chemical products, this leads mainly to agreement on the conditions of putting a product on the market, and agreement on the industrial conditions of product use.

A3.3 Agreement on the Conditions for Market Approval

It is important to distinguish between directives aimed at pure substances and those aimed at their mixtures, otherwise known as *preparations*. Obviously, it is the pure substances that were the first to be the object of specific EC measures; their number, however large, is actually limited compared to the theoretically infinite variety of their mixtures.

A3.3.1 Substances

The first EC text on the subject of substances was Directive 67/548/CEE of 27 June 1967 regarding the classification, packaging, and labeling of dangerous substances.

Since its adoption, this directive has been modified 7 times by the EEC Council, and updated to match technical progress 16 times by the EC Commission. The next-to-last modification, by Directive 79/831/CEE of 18 September 1979, led to the introduction in French law of a preliminary notification procedure for new substances (Decree 86-570 of 14 March 1986, Articles R231-51 and following, of the Labor Code) and to the complete revision of the rules governing the labeling and packing of pure substances (modification of Article L231-6 of the Labor Code, decree of 10 October 1983).

The latest modification (Directive 92/32/CEE of 30 April 1992) as well as the sixteenth amendment (Directive 92/37/CEE of 30 April 1992) have not yet been entered into French law as of October 1993. Taking into account the experience gained since 1979, and the practical difficulties encountered by most member states following the sixth modification, the seventh modification, of 30 April 1992, introduced sensible amendments to many parts of the directive. These include changes to the present procedures of market approval for new substances, to the definitions of those considered dangerous, to the makers' obligations concerning research and furnishing of toxicological data, to the rules of preservation of confidentiality of certain information, and finally, to the data to be transmitted to the users by the indirect means of an index of safety data. On this last point, the directive of 30 April, 1992 makes the directive of 7 June 1988 on preparations applicable for substances that have already been instituted, namely, the indexes of safety data.

The successive changes to Directive 67-548 consist essentially of introducing an appendix of new substances to be labeled and of specifying the nature of certain toxicologic tests necessary before a substance may go on the market (decree of 14 April 1989 modifying the decree of 14 March 1986 regarding information and test results to be furnished as required in application of Article R231-1 of the Labor Code). It is also in this way that the decree of 10 October 1983 is regularly updated (latest modification: decree of 16 January 1992).

A3.3.2 Preparations

Concerning chemical preparations, it is important to distinguish between two stages characterized by two appreciably different steps:

- –First, the EEC institutions are limited to regulations on the market approval of certain categories of preparations considered either the most dangerous or the most important from an economic standpoint;
- –Second, all dangerous preparations are accounted for, that is, all those for which the member states have general basic regulations restricting the sale or use.

Regarding those texts whose transferral to internal law is inscribed in the Labor Code, excluding medications and certain phytosanitary products for which specific regulations already exist, these are covered by the first stage:

- –Directive 73/173/CEE of 4 June 1973 concerning solvents (modified 22 November 1980 by Directive 80/781/CEE and amended 10 June 1982 by Directive 82/473/CEE); this directive was reversed by the decree of 1983, now repealed;
- –Directive 77/728/CEE of 7 November 1977 concerning paints (modified 16 May 1983 by Directive 83/265/CEE and changed twice, most recently by Directive 86/508/CEE of 7 October 1986); this directive was reversed by the decree of 7 October 1983, now repealed (Article 28 of the decree of 21 February 1990).

Since June 1991, the so-called all preparations directive of 7 June 1988 (88/739/CEE) has permitted the preliminary classification for their market approval, and above all, the labeling of the whole of the dangerous preparations; with this end in view, conventional methods of classification based on respective quantities of dangerous substances present in a preparation have been broadened. Negotiations on this point were obviously long in coming to a common positon among the member states in December 1987, and again to its adoption on 1988. A decree of 21 February 1990 transferred the essence of the directive of 7 June 1988 to French law. This decree was itself modified in January 1992 to take into account the first amendments to the 1988 directive, notably concerning dangerous preparations containing methanol or dichloromethane.

A3.4 Agreement on Conditions of Product Use

This section deals with the coming to an agreement within the industry on the conditions of hygiene and safety, and the conditions of hygiene and work: the directives aim to impose certain rules on the employer concerning minimum safety requirements, which are based on Article 118A of the Unification Act of 1987.

The foundation of the text in this area is the directive outline of 27 November 1980 (80/1107/CEE) modified by a directive of 16 December 1988 setting the general principles of the regulations regarding safety and hygiene in the use of physical, chemical, or biological agents.

The principles developed by this directive are as follows:

–the obligation of respecting (TLVs) in the workplace;
–regular testing of work areas according to standardized procedures;
–observation of the state of health of exposed workers.

Several directives applicable to the directive of 27 November 1987 have been adopted that tend to specify certain minimum regulations for a particular risk, particularly the TLVs. These include:

–Directive 78/610/CEE of 29 June 1978 on vinyl chloride, which was the subject of a decree of transferral in 1980;
–Directive 82/605/CEE of 28 July 1982 on lead, transferred by a decree of 1 February 1988;
–a Noise Directive of 12 May 1986, transferred by the decree of 21 April 1988 (Articles R232-8ff. of the Labor Code);
–a directive of 9 June 1988 (88/364/CEE) prohibiting certain carcinogenic products: four aromatic amines. This document was actually transferred to French law by Decree 89/593 of 28 August 1989, which only authorized their use for purposes of research testing, scientific analysis, or waste elimination;
–a directive of 28 June 1990 (90/394/CEE) "on carcinogenic products handled in a professional environment" that is in the process of transferral;
–Directive 90/679/CEE of 26 November 1990 on the protection of workers exposed to biological agents, also in the process of transferral.

Finally, it is important to point out that the document does not specifically address the problem of toxic risks. This directive of 12 June 1989 "concerning the enactment of measures aimed at promoting the betterment of the health and safety of the workers," called a "new directive outline" defines, in a general way, the respective obligations of employers and employees in the area of prevention, the procedures of information and dialogue with the workers, and it introduces to the EC level the idea of monitoring the state of health of all the workers as well as that of protection and prevention, as, for example, the French labor medical services already do. One must also cite the "SEVESO"

Directive 82/501/CEE of 24 June 1982, whose enactment comes mainly from the regulations on classified establishments.

A3.5 Procedure for the Drawing Up of European Directives in the Domain of Chemical Products

A3.5.1 Proposal by the European Commission

The ECC, which is an autonomous institution of the community that, like the European Parliament, the Court of Justice at Luxemburg, or the European Council, has the job of proposing directives; during the different phases of setting up projects, it consults, formally or informally, with groups of experts, and the consulting committee for safety, hygiene, and health protection in the workplace (made up of government representatives and the national social partners).

A3.5.2 Consultation of the Economic and Social Committee and of the European Parliament

Before submitting their project to the Council of Communities, the EC Commission must obtain the opinion of the Economic and Social Committee, and of the European Parliament.

A3.5.3 Tabling of the Proposals by the Council

The Council, has the European Commission's directive proposal studied by The Social Questions Group or the Economic Questions Group, and the Committee of Permanent Representatives of the States.

A3.5.4 Adoption by the Council of Labor Ministers

The directives are adopted by the Council of Labor Ministers first, by informal vote on the majority decision, and second, after the second reading in the European Parliament of the definitive document.

A3.5.5 Notification of the Directive to the Member States

Directives are not directly applicable in the member states. They must be notified on the directives, as they are the sole recipients, and have the task of transferring the EC directives to their national laws.

The principle is that, because of the differences in European regulations, the directives are the final authority, but the job of determining the necessary methods of implementation should be left to the states' national authorities.

A3.6 "Biological" Directives

Predating national regulations for the most part, the EEC has been engaged for close to 5 years in drawing up specific provisions for dealing with the risks associated with biotechnology.

Regarding techniques that could have an effect on the environment as well as on people exposed to biological agents as part of their profession, the two directives presently adopted by the EC Council and henceforth proposed to the states are based, of course, on two articles of the Treaty of Rome:

- —Article 1305, as it concerns the directive of 23 April 1990, relates to the restricted use of genetically modified microorganisms,
- —Article 118A, concerning the directive ratified by the Community Council of Labor Ministers on 26 November 1990, and which has to do with "the protection of workers from the risks associated with exposure to biological agents at work."

This two-pronged legal approach shows the EC Council's desire to deal with all these problems: on the one hand, that of genetically modified microorganisms, ensuring the protection of the environment and taking every precaution to avoid the spread of pathogenic microorganisms, and on the other hand, promoting the workers' protections.

Article 1305 of the treaty, introduced by the Unification Act, permits, in effect, the EC Council to take any necessary measures to preserve, protect, and improve the quality of the environment. This article has, so far, not had nearly as many regulatory applications as Article 118A. With this in mind, it is important to note that the "Workers' Directive" of 26 November 1990 is the seventh application of the General Directive 89/391/CEE of 12 June 1989.

Finally, before briefly defining the scope of the two EC texts in the domain of biotechnology, it is necessary to emphasize that the directive of 23 April 1990, which applies only to genetically modified microorganisms, necessarily has a more limited scope than the directive of 26 November 1990, where the field of application includes all pathogenic biological agents, whether they are the result of a genetic mutation, or a cellular culture, or whether they are human endoparasites suspected of causing infections, allergies, or poisoning.

A3.6.1 The Directive of 26 April 1990 Relative to the Restricted Use of Genetically Modified Microorganisms

The general structure of the directive of 26 April 1990 is to protect the environment and to ensure, when opportune, the rigorous supervision of activities seeking to modify genetic materials by national authorities.

Of course, the directive defines what it means by "genetically modified microorganisms," excluding modifications of natural mutations such as in vitro fertilization, polyploid induction, or conjugation, as these changes do not employ the techniques of DNA recombination.

Microorganisms are divided in two groups according to their pathogenic characteristics. Moreover, the directive distinguishes between the activities of teaching and research, and, in a general way, between those that are not designed for industrial use, and those that are especially designed for commercial production.

Taking these elements into account, the users or promoters of genetically modified microorganisms are required, on the one hand, to evaluate the risk of contamination and implement the measures of restriction that are imposed, and on the other hand, to declare or apply for a preliminary authorization for its use. Naturally, employers must inform their personnel and establish an emergency plan in case of accident before the start of each operation.

A3.6.2 Directive of 26 November 1990

The directive of 26 November 1990 sets the principles of prevention, ensuring an effective protection against eventual infection of the workers most likely to be exposed to pathogenic biological agents. To this end, a notice is to be placed in all relevant work areas that includes instructions for a containment system for microorganims appropriate for the nature of the dangers and procedures taking place. The implementation of these measures obviously assumes that the risks were correctly evaluated as much from the viewpoint of what products are being handled as from the number and positions on the workers concerned.

This directive applies to all businesses in which biological agents (fermentation, chemical synthesis, genetics, medicine) are used in one form or another, but for reasons inherent to their use, different systems are planned respectively for the medical services, diagnostic laboratories, and industrial establishments.

Biological agents are divided into four groups, according to how much of a risk of infection they present.

On receipt of the risk evaluation, which must be done in all cases, and which, if necessary, must be periodically renewed, the employer takes the appropriate collective and individual protective measures. These various measures must also be brought to the attention of the administrative authorities, as must the list of endangered workers. Notice must be given to the administration before the use of the most dangerous biological agents (groups 2, 3, and 4).

Finally, the directive requires specific medical supervision for workers likely to be exposed to these agents in the course of their duties.

As for medical and veterinary services other than those of the diagnostic laboratories, the directive requires the enactment of specific measures that will take into account the uncertainty as to the nature and effective presence of pathogenic biological agents.

In conclusion, this directive opens a hitherto unexplored area of French law and ambitiously so; since despite the difficulties inherent in the diversity of activities involving exposure to biological agents and the multiplicity of the circumstances of exposure, the EEC document tries to establish equally protective regulations for all workers. If only because of these two directives, the social agenda of the community is not negligible since it is not limited to repeating the existing national systems, but on the contrary, leads the member states to take into account the areas where the risks are not yet specifically regulated.

More on the Biological Risks: The Prions

There are some infectious diseases about which we understand neither the origin nor the exact nature of their causal agent, which is neither a bacterium nor a classic virus. They are sometimes called *slow viruses* or *viroids*. At the top of the list of infectious attackers of unknown origin are the transmissible spongiform encephalopathies (TSE).

Among these maladies, some affect humans: this is the case for the Creutzfeldt–Jakob disease, which is generally transmitted by imperfectly sterilized surgical instruments, by origin transplants from afflicted donors (corneal grafts), or by growth hormone treatments originating from human pituitary glands (10 cases in children described in France since 1988). The sheep-shaking sickness (Scrapie) or the mad cow sickness may attack other mammals.

Whether they concern animal or human ailments, these neurodegenerative encephalopathies have as a common infectious agent a protinaceous infectious particle, which Prusiner (1982)* named a *prion* (or PrP particle). It consists of a glycoprotein (PM = 27 Kdal) without its nucleic acid.[†]

This particular protein is extremely resistant to heat (sterilizer or dry heat at 160°C or more), to ultraviolet (UV) radiation, and to bridging reagents such as formaldehyde or glutaraldehyde; however, it is susceptible to phenol and sodium hypochlorite.

---

*S. B. Prusiner. Novel protinaceous infectious particle causes Scrapie. *Science.* 1982, *198*, 136–44.

[†]J. P. Liautard. Les prions sont-ils des molécules chaperonnes mal repliées? *Medecine Sciences 8*, 1.

The apparently easy transmission between species and the resistance to classic denaturation treatments (heat, UV, aldehydes) must inspire the maximum respect for safety procedures (protective glasses, disposable gloves, etc.) in all who handle any of the nervous tissues of animals or humans.

All histological materials, including fixed samples, must always be considered extremely infectious.* All refuse and contaminated single-use materials must be sterilized before being destroyed in impervious containers in an incinerator licensed for special wastes.†

The present uncertainty about the exact nature of the prion and about its actual mode of transmission cannot help but cause great caution, as is the case for all as-yet ill-defined biological risks.

---

*P. Brown. The effect of chemical, heat, and histopathologic processing on high-infectivity hamster-adapted Scrapie virus; Laboratory and hospital disinfection of spongiform encephalopathy viruses. In: L. A. Court, ed. *Virus non conventionnels et affectations du SNC*, Masson, Paris, pp. 156–63; 510–17.

†J. Paul. Attention: Prion! *Inserm-Actualités*, no. 105, April 1992, pp. 7–8.

Bibliography

Chemical Risks

General References

N. V. Steere. *Safety in the chemical laboratory.* Journal of Chemical Education, Easton (USA), Vol. 1 (1967), Vol. 2 (1971), Vol. 3 (1978).

P. J. Gaston. *The care, handling and disposal of dangerous chemicals.* Northern Publishers, Aberdeen (1970).

_____. *Safety in the science laboratory and in the use of chemicals.* Imperial College of Science and Technology, London (1970).

_____. *Guide for safety in the chemical laboratory.* 2nd Ed. Manufacturing Chemists Association, Van Nostrand Reinhold Co., New York (1972).

H. A. J. Pieters and J. W. Creyghton. *Safety in the chemical laboratory.* 2nd Ed. Butterworths, London (1975).

_____. *Safety in academic chemistry laboratories.* 5th Ed. American Chemical Society, Washington (1991).

E. Meyer. *Chemistry of hazardous materials.* 2nd Ed. Prentice-Hall, Englewood Cliffs (1989).

J. G. Ellis and N. J. Riches. *Safety and laboratory practice.* Macmillan Press, London (1978).

N. E. Green and A. Turk. *Safety in working with chemicals.* Macmillan Press, New York (1978).

J. A. Kaufman. *Laboratory safety guidelines.* Wellesley (1978).

N. H. Proctor and J. P. Hughes. *Chemical hazards in the workplace.* 3rd Ed. Van Nostrand Reinhold, New York (1991).

K. M. Reese. *Health and safety guidelines for chemistry teachers.* American Chemical Society, Washington (1979).

M. M. Renfrew. *Safety in the chemical laboratory.* Journal of Chemical Education, Easton (USA), Vol. 4 (1981).

D. B. Walters and C. W. James. *Health and safety for toxicity testing.* Butterworths (1984).

L. J. Diberardinis, J. Baunet, M. W. First, G. T. Gatwood, E. Groden, and A. Seth. *Guidelines for laboratory design. Health and safety considerations.* Wiley Interscience, New York.

V. C. Marshall. *Major chemical hazards.* John Wiley and Sons, New York (1987).

J.P. Dux and R. F. Stalzer. *Managing safety in the chemical laboratory.* Van Nostrand Reinhold, New York (1988).

———. *Safety in the school science laboratory.* DHEW, Council of State Science Supervisors, Richmond (1979).

———. *Handling chemicals safely.* 2nd Ed. Dutch Association of Safety Experts, Dutch Chemical Association, Dutch Safety Institute, Amsterdam (1980).

D. B. Walters. *Safe handling of chemical carcinogens, mutagens, teratogens and highly toxic substances.* Vol. 1, Vol. 2. Ann Arbor Science Pub., Ann Arbor (1980).

A. Fuscaldo, B. J. Erlich and B. Hindman. *Laboratory safety, theory and practice.* Academic Press, New York (1980).

G. Choudary. *Chemical hazards in the workplace: measurement and control.* American Chemical Society, Washington, ACS Symposium Series no. 149 (1981).

———. *Hazardous chemicals: a manual for schools and colleges.* 4th Ed. Olivier and Boyd, Edinburgh (1981).

———. *Prudent practices for handling hazardous chemicals in laboratories.* 5th Ed. National Academic Press, Washington (1991).

H. Fawcett and W. S. Wood. *Safety and accident prevention in chemical operations.* 2nd Ed. John Wiley and Sons, New York (1982).

N. T. Freeman and J. Whitehead. *Introduction to safety in the chemical laboratory.* Academic Press, New York (1982).

J. T. Snow. *Handling of carcinogens and hazardous compounds.* Calbiochem-Behring, San Diego (1982).

———. *Health and safety in the chemical laboratory, where do we go from here?* No. 51. The Royal Society of Chemistry, London (1987).

———. *La sécurité avec Merck.* Edn Merck, Darmstadt (1983).

———. *La sécurité dans les laboratoires utilisant des substances chimiques.* Centre de Prévention et de Protection, Paris (1983).

———. *Prudent practices for the disposal of hazardous chemicals from the laboratory.* National Academic Press, Washington (1983).

M. L. Richardson. *Risk assessment of chemicals in the environment.* Royal Society of Chemistry, London (1989).

R. J. Lewis. *Rapid guide to hazardous chemicals in the workplace.* 2nd Ed. Chapman Hall, London (1990).

R. Scott. *Chemical hazards in the workplace.* Lewis Pub., Chelsea (1990).

R. Lewis. *Hazardous chemical desk reference.* 2nd Ed. Van Nostrand Reinhold, New York (1991).

W. Mahn. *Academic laboratory chemical hazards guidebook.* 2nd Ed. (1991).

S. G. Luxon. *Hazards in the chemical laboratory.* Royal Society of Chemistry, London (1992).

B. Montfort. *La sécurité dans les laboratoires d'enseignement de chimie.* Société française de Chimie, Paris (1992).

H. H. Fawcett. *Hazardous and toxic materials: safe handling and disposal.* 2nd Ed. John Wiley and Sons, New York (1988).

D. A. Pipitone. *Safe storage of laboratory chemicals.* 2nd Ed. John Wiley and Sons, New York (1991).

———. *School science laboratories. A guide to some hazardous substances.* Council of State Science Supervisors, Richmond (1984).

L. Bretherick. *Hazards in the chemical laboratory.* 4th Ed. Royal Society of Chemistry, London (1986).

———. *Guide to safe practices in chemical laboratories.* Royal Society of Chemistry, London (1986).

Y. Yoshida. *Safety reactive chemicals.* Industrial safety series 1. Elsevier Applied Science Pub., Barking (1986).

S. Lipton and J. Lynch. *Health hazard control in the chemical process industry* (1987).

J. A. Young. *Improving safety in the chemical laboratory: a practical guide.* 2nd Ed. John Wiley and Sons, Chichester (1991).

M. A. Armour. *Hazardous laboratory chemicals: disposal guide.* CRC Press, Boca Raton (1991).

Dictionaries, Handbooks

Toxic and hazardous industrial chemicals : safety manual for handling and disposal. Toxicity and hazard data. The International Technical Information Institute, Japan (1988).

A. K. Furr. *CRC Handbook of laboratory safety.* 3rd Ed. CRC Press Inc., Boca Raton (1989).

H. Sittig. *Hazardous and toxic effects of industrial chemicals.* Noyes Data Corporation, Park Ridge (1981).

R. Roi, W. G. Town, W. G. Hunter and L. Alessio. *Occupational health guidelines for chemical risk.* Commission of the European Communities, Luxembourg (1983).

P. Howard. *Handbook of environmental fate and exposure data for organic chemicals.* Vol. I: *Large production and priority pollutants* (1989); Vol. II: *Solvents* (1990); Vol. III : *Pesticides* (1990). Royal Society of Chemistry, London.

R. Lewis. *Sax's dangerous properties of industrial materials.* 8th Ed. Van Nostrand Reinhold, New York (1992).

M. Richardson and S. Gangoli. *Dictionary of substances and their effects.* (5 volumes) Vol. 1 (A–C), Royal Society of Chemistry, London (1992).

S. Bubavari. *The Merck index.* 11th Ed. Merck and Co., Rahway (1989).

N. I. Sax. *Dangerous properties of industrial materials.* 7th Ed. Van Nostrand Reinhold Co., New York (1988).

L. H. Keith and D. B. Walters. *Compendium of safety data sheets for research and industrial chemicals.* VCH Publishers Inc., New York, Vols. 1–3 (1985), Vols. 4–6 (1987) ; Vol. 7, L. H. Keith, D. B. Walters and T. C. Zebovitz (1989).

———. *The national toxicology program's chemical data compendium.* Vols. I–VIII, Lewis Pub. (1992).

———. *The national toxicology program's chemical solubility compendium.* Lewis Pub. (1992).

H. Sittig. *Handbook of toxic and hazardous chemicals and carcinogens.* 4th Ed. Noyes Pub., Park Ridge (1991).

N. I. Sax. *Hazardous chemicals information.* Annual No. 1. Van Nostrand Reinhold Information Services, New York (1986).

N. I. Sax and R. J. Lewis. *Rapid guide to hazardous chemicals in the workplace.* 2nd Ed. Van Nostrand Reinhold Co., New York (1990).

G. Weiss. *Hazardous chemicals data books.* 2nd Ed. Noyes Data Corporation, Park Ridge (1986).

NIOSH pocket guide to chemical hazards. DHHS (NIOSH) Publication No. 90-117, NIOSH, Cincinnati (1990).

E. R. Plunkett. *Handbook of industrial toxicology.* 3rd Ed. Chemical Publishing Co., New York (1987).

M. Richardson. *Toxic hazard assessment of chemicals* Royal Society of Chemistry, Nottingham (1987).

N.I. Sax. *Dangerous properties of industrial materials: report.* Vol. 1. Van Nostrand Reinhold Co., New York, regular publication (1987).

R. S. Stricoff and D. B. Walters. *Laboratory health and safety handbook: a guide to the preparation of a chemical hygiene plan,* John Wiley, New York (1990).

Cryogenic Fluids

M. G. Zabetakis. *Safety with cryogenic fluids.* Haywood Books, London (1967).

———. *Cryogenic safety manual. A guide to good practice.* Mech Eng. Pub., GB (1982).

F. D. Eskuty and K. D. Williams. *Liquid cryogens.* 2nd Ed. CRC Press, Boca Raton (1983).

Compressed Gas

Encyclopedia of gases. Elsevier, Liquid Air, Amsterdam (1976).
Handbook of compressed gases. Compressed gas associated. 3rd Ed. Van Nostrand Reinhold, New York (1990).

Guide to safe handling of compressed gases. Matheson Publications, Hasbrouck Heights, N.J. (1990).

W. Braker and A. L. Mossman. *Effects of exposure to toxic gases: first aid and medical treatment.* 2nd Ed. Matheson Publication, Secaucus (1977).

C. F. Cullis and J. G. Firth. *Detection and measurement of hazardous gases.* Heinemann, London (1981).

W. Braker and A. L. Mossman. *The Matheson gas data book.* 6th Ed. Matheson Gas Products, Secaucus (1981).

History of Chemical Accidents

Case histories of accidents in the chemical industry. Monthly Publication, MCA, Washington (1953–1975 and 1978).

A. J. Smith, Jr. *Managing hazardous substances accidents.* McGraw-Hill, New York (1981).

Explosive Substances

B. T. Federoff. *Encyclopedia of explosives and related compounds.* Dover, N.H. Piccatinny Arsenal. Vol. 1 (1960), Vol. 10 (1984).

T. Urbanski. *Chemistry and technology of explosives.* Pergamon Press, Oxford. Vol. 1 (1964), Vol. 2 (1965), Vol. 3 (1967), Vol. 4 (1984).

J. Calzia. *Les substances explosives et leurs nuisances.* Dunod, Paris (1969).

_____. *Les mélanges explosifs, gaz et vapeurs, nuages de poussières, liquides et solides.* INRS, Paris (1976).

D. Stull. *Fundamentals of fire and explosion.* Monograph series no. 10, American Institute of Chemical Engineers, New York (1977).

L. Menard. *Les explosifs occasionnels.* Vols. 1 and 2. Techniques et Documentation, Paris (1987).

Inflammable Substances

C. W. Bahme. *Fire protection for chemicals.* National Fire Protection Association International (1961).

G. P. McKinnon. *Fire protection handbook.* 14th Ed. National Fire Protection Association, Quincy (1976).

A. Vallaud and R. Damel. *Incendie et produits chimiques.* Société Alpine de Publication, Grenoble (1984).

A. H. Landrock. *Handbook of plastics: flammability and combustion toxicology.* Noyes Pub., Ridge Park (1983).

Oxidizing Agents

J. C. Schumacher. *Perchlorates, their properties, manufacture and uses.* ACS no. 146, Reinhold, New York (1960).

E. W. Lawless and I. C. Smith. *Inorganic high-energy oxidizers.* Marcel Dekker Inc., New York (1968).

H. M. Castrantas and D. K. Banerjee. *Laboratory handling and storage of peroxy compounds.* ASTM, Philadelphia (1970).

D. Swern. *Organic peroxides.* Wiley Interscience, London. Vol. 1 (1970), Vol. 2 (1971), Vol. 3 (1972).

Reactive Substances and Dangerous Chemical Reactions

D. R. Cloyd. *Handling hazardous materials.* National Aeronautics and Space Administration, Washington (1965).

J. R. Gibson. *Handbook of selected properties of air and water reactive materials.* U.S. Naval Munition Depot, Crane, Indiana (1969).

_____. *Hazardous chemicals data.* National Fire Protection Association, Boston (1971).

_____. *Manual of hazardous chemical reactions.* 5th Ed. No. 491M. National Fire Protection Association, Boston (1975).

T. Yoshida. *Handbook of hazardous reactions with chemicals.* Tokyo Fire Department, Tokyo (1980).

L. Bretherick. *Handbook of reactive chemical hazards.* 4th Ed. Butterworths, London (1990).

D. F. Shriver and M. A. Drezdzon. *The manipulation of air sensitive compounds.* 2nd Ed. Wiley Interscience, New York (1986).

J. Leleu. *Réactions chimiques dangereuses.* INRS, Paris (1987).

Regulations and Threshold Values

G. S. Dominguez and K. G. Bartlett. *Hazardous waste management.* Vol. 1 : The law of toxics and toxic substances. CRC Press Inc, Boca Raton (1986).

_____. *Substances et préparations chimiques dangereuses: conditions d'étiquetage et d'emballage.* Ministère des Affaires Sociales et de l'Emploi, Paris (1990).

J. P. Pluyette. *Hygiène et sécurité. Conditions de travail, lois et textes réglementaires.* 20ème Ed. Technique et Documentation, Lavoisier, Paris (1991).

_____. *Maximum concentrations at the work place and biological tolerance values for working materials.* Report no. 23, Commission for the Investigation of Health Hazards of Chemical Compounds in the Work Area, Weinheim (1987).

_____. *Threshold limit values and biological exposure indices for* 1991–1992, ACGIH, Cincinnati.

_____. *Valeurs limites pour les concentrations des substances dangereuses dans l'air des lieux de travail.* INRS, Paris, annual update.

_____. *Valeurs seuil limites.* ICF, Bruxelles (1992).

_____. *Occupational toxicants: critical data evaluation for MAK values and classification of carcinogens.* Vol. 3. VCH, Weinheim (1992).

Techniques of Destruction

E. Ellsworth Hackmann III. *Toxic organic chemicals: destruction and waste treatment.* Noyes Data Corporation, Park Ridge (1978).

_____. *Prudent practices for disposal of chemicals from laboratories.* National Academy Press, Washington (1983).

_____. *Treatment and disposal methods for waste chemicals.* International Register of Potential Toxic Chemicals, United Nations Environment Program, Geneva (1985).

M. J. Pitt and E. Pitt. *Handbook of laboratory waste disposal.* Ellis Horwood Limited, Chichester (1986).

_____. *Laboratory waste disposal manual.* 2nd Ed. Manufacturing Chemists Association, Washington (1975).

_____. *The disposal of hazardous waste from laboratories.* The Royal Society of Chemistry, London (1983).

G. Lunn and E. B. Sansone. *Destruction of hazardous chemicals in the laboratory.* John Wiley and Sons, New York (1990).

M. A. Armour. *Hazardous laboratory chemicals disposal guide.* CRC Press, Boca Raton (1991).

M. Maes. *La maîtrise des déchets industriels.* Pierre Johanet S.A., Paris (1991).

Toxicology

General References

R. Fabre and R. Truhaut. *Précis de toxicologie.* Vol. 1, Vol. 2 Sédes Ed., Paris (1965).

E. J. Ariens, A. M. Simonis and J. Offermeier. *Introduction to general toxicology.* Academic Press, New York (1976).

Patty's industrial hygiene and toxicology John Wiley and Sons, New York.
Vol. 1. G. D. Clayton and F. E. Clayton. *General principles.* 4th revised Edition (1991). Part A and B.
  Vol. 2A. G. D. Clayton and F. E. Clayton. *Toxicology.* 3rd revised Edition (1981).
  Vol. 2B. G. D. Clayton and F. E. Clayton. *Toxicology.* 3rd revised Edition (1981).
  Vol. 2C. G. D. Clayton and F. E. Clayton. *Toxicology.* 3rd revised Edition (1981).
  Vol. 3. L. R. Cralley and L. V. Cralley. *Theory and rationale of industrial hygiene practice.* 2nd Ed. Part A and B (1985).

T. A. Loomis. *Essentials of toxicology.* 3rd Ed. Lea & Febiger, Philadelphia (1978).

A. Picot. *Aspect biochimique de la toxicité de diverses substances chimiques (solvants, substances mutagènes, cancérogènes).* CNRS, Gif-sur-Yvette (1979).

J.M. Haguenoer and D. Furon. *Toxicologie et hygiène industrielle.* Technique et Documentation, Lavoisier, Paris.
  Vol. 1. *Les dérivés minéraux, première partie* (1981).
  Vol. 2. *Les dérivés minéraux, deuxième partie* (1982).
  Vol. 10. *Le comportement en milieu de travail* (1984)

A. Reeves. *Toxicology, principles and practice.* Vol. 1 John Wiley and Sons, New York (1981).

R. L. Lauwerys. *Toxicologie industrielle et intoxications professionnelles.* 2ème Ed. Masson, Paris (1990).

A. Tu. *Survey of contemporary toxicology.* John Wiley and Sons, New York. Vol. 1 (1980), Vol. 2 (1982).

L. Goldberg. *Structure activity correlation as a predictive tool in toxicology: fundamentals, methods and applications.* Hemisphere Pub., Washington (1983).

E. Hodgson and F. E Guthrie. *Introduction to biochemical toxicology.* 2nd Ed. Elsevier, New York (1986).

T. J. Haley and W. O. Berndt. *Toxicology.* Hemisphere Pub., Washington (1987).

―――. *Handbook of toxicology.* Taylor and Francis, London (1988).

M. A. Kamrin. *Toxicology: a primer on toxicology principles and applications.* Royal Society of Chemistry, Cambridge (1988).

S. E. Manahan. *Toxicological chemistry. A guide to toxic substances in chemistry.* Lewis Pub., Chelsea (1989).

J. K. Marquis. *Guide to general toxicology.* 2nd Ed. Karger, Basel (1989).

J. A. Timbrell. *Introduction to toxicology.* Taylor and Francis Ltd., London (1989).

E. Hodgson, R. B. Mailman and J. E. Chambers. *Dictionary of toxicology.* Van Nostrand Reinhold, New York (1990).

G. N. Volans, J. Sims, F. M. Sullivan, P. Turner. *Basic science in toxicology.* Taylor and Francis Ltd., London (1990).

J. A. Timbrell. *Principle of biochemical toxicology.* Taylor and Francis Ltd., London (1982).

F. Sperling. *Toxicology, principles and practice.* Vol. 2. John Wiley and Sons, New York (1984).

A. Albert. *Selective toxicology: the physico-chemical basis of therapy.* 7th Ed. Chapman and Hall, London (1985).

R. E. Gosselin, H. C. Hodge, R. P. Smith and M. N. Gleason. *Clinical toxicology of commercial products, acute poisoning*. 5th Ed. Williams & Wilkins, Baltimore (1985).

———. *Objectifs des essais toxicologiques*. Chimie et Ecologie, Paris (1985).

M. O. Amdur, J. Doull and C. D. Klassen. *Casarett and Doull's toxicology. The basic science of poisons*. 4th Ed. Pergamon Press, New York (1991).

C. Bismuth, F. Baud, F. Conso, J. P. Frejaville and R. Garnier. *Toxicologie clinique*. 4th Ed. Flammarion Médecine Sciences, Paris (1987).

A. W. Hayes. *Principles and methods of toxicology*. 2nd Ed. Raven Press, New York (1987).

E. Hodgson and P. E. Levi. *A textbook of modern toxicology*. Elsevier, New York (1987).

F. C. Lu. *Basic toxicology fundamentals. Target organs and risk assessment*. 2nd Ed. Hemisphere Pub., Washington (1991).

Toxicology of Mineral Compounds

L. Friberg, G. F. Nordberg and V. B. Vouk. *Handbook on the toxicology of metals*. Elsevier, Amsterdam (1979).

H. Sigel. *Metal ions in biological systems*. Vol. 20: Concepts on metal ion toxicity. Marcel Dekker, New York (1986).

H. G. Seiler and H. Sigel. *Handbook on toxicity of inorganic compounds*. Marcel Dekker, New York (1987).

G. E. Browning. *Toxicity of industrial metals*. 2nd Ed. Butterworths, London (1969).

S. S. Brown. *Clinical chemistry and chemical toxicology of metals*. Elsevier Scientific Pub., Amsterdam (1977).

R. A. Goyer and M. A. Mehlan. *Toxicology of trace metals*. John Wiley and Sons, New York (1977).

E. Merian, R. W. Frei, W. Hardt, C. Schlatter. *Carcinogenic and mutagenic metal compounds*. Gordon and Breach Science Pub., New York (1985).

P. J. Craig. *Organometallic compounds in the environment*. John Wiley and Sons, New York (1986).

G. L. Fischer and M. A. Gallo. *Asbestos toxicity*. Marcel Dekker, New York (1988).

Toxicology of Solvents

E. Browning. *Toxicity and metabolism of industrial solvents*. Elsevier, New York (1965).

A. J. Collings and S. G. Luxon. *Safe use of solvents*. Academic Press, London (1982).

A. Englund, K. Ringen and M. Mehlan. *Occupational health of solvents*. Princeton Scient. Pub., Princeton (1982).

G. Kakabadse. *Solvent problems in industry*. Elsevier, London (1984).

D. J. De Renzo. *Solvents safety handbook*. Noyes Data Corp., Park Ridge (1986).

———. *Organo-chlorine solvents. Health risks to workers*. Commission of the European Communities, Royal Society of Chemistry, Brussels, Luxembourg (1986).

R. Snyder. *Ethel Browning's toxicity and metabolism of industrial solvents*. 2nd Ed., Vol. I : *Hydrocarbons* (1987). Vol. II: *Nitrogen and phosphorus solvents* (1989). Elsevier, Amsterdam.

V. Rihimalk and U. Ulfvarson. *Safety and health aspects of organic solvents*. Progress in clinical and biological research. Vol. 220 A. R. Liss Inc., New York (1986).

Genetic Toxicology : Mutagenesis and Carcinogenesis

L. Fishbein. *Potential industrial carcinogens and mutagens.* Elsevier, Amsterdam (1979).

Haguenoer, P. Frimat, J. Bonneter and Ph. Vennin. *Les cancers professionnels.* Technique et Documentation, Lavoisier, Paris (1982).

R. F. Deisler. *Reducing the carcinogenic risks in industry.* Marcel Dekker, New York (1984).

M. Hofnung. *Tests à court terme en toxicologie génétique. Principes et techniques.* Les Editions de l'INSERM, Paris (1984).

C. E. Searle. *Chemical carcinogens.* Vol. 1 and 2. ACS Monograph 182, 2nd Ed. American Chemical Society, Washington (1984).

J. Moutschen. *Introduction to genetic toxicology.* John Wiley and Sons, New York (1984).

M. Castegnaro and E. B. Sansone. *Chemical carcinogens: some guidelines for handling and disposal in the laboratory.* Springer-Verlag, Berlin (1986).

M. Kirsch-Volders. *Mutagenicity, carcinogenicity and teratogenicity of industrial pollutants.* Plenum Press, New York (1984).

———. *Measurement techniques for carcinogenic agents in the workplace air.* Royal Society of Chemistry, Cambridge (1989).

A. Berlin, M. Draper, E. Kaug, R. Roi and M. T. Van Der Venne. *The toxicology of chemicals: Part I. Carcinogenicity:* Vol. 1 (1989), Vol. 2 (1990). Commission of European Communities, Luxembourg.

R.J. Lewis. *Carcinogenically active chemicals. A reference guide.* Van Nostrand Reinhold, New York (1990).

———. *Reproductively active chemicals. A reference guide.* Van Nostrand Reinhold, New York (1991).

N. Bichet, P. Boutibonnes, Y. Courtois, J. Dayan, D. Marzin, N. Mazaleyrat, and N. Weill. *Manipulation des mutagènes et descancérogènes. Guide de bonnes pratiques de sécurité,* Société Française de Toxicologie Génétique, Paris (1990).

Industrial Toxicology

A. J. Finkel. *Hamilton and Hardy's industrial toxicology.* 4th Ed. John Wright, Boston (1983).

E.R. Plunkett. *Handbook of industrial toxicology.* 3rd Ed. Chemical Pub. Co., New York (1987).

R. Lauwerys. *Industrial chemical exposure: guidelines for biological monitoring.* Biomedical Publications (1983).

N. K. Woodward. *Phthalate esters: toxicity and metabolism.* Vols. I and II. CRC Press, Boca Raton (1988).

P. L. Williams and J. L. Burson. *Industrial toxicology, safety and health applications in the workplace.* Van Nostrand Reinhold Co., New York (1985).

R. Lauwerys. *Toxicologie industrielle.* Masson, Paris (1990).

V. O. Sheftel. *Toxic properties of polymers and additives.* Rapa Ed., Shawbury (1990).

Molecular Toxicology

D. E. Hathway. *Molecular aspects of toxicology.* The Royal Society of Chemistry, London (1984).

F. De Matteis and E. A. Lock. *Selectivity and molecular mechanisms of toxicity.* Macmillan Press, London (1987).

L. J. Mamett. *Frontiers in molecular toxicology.* American Chemical Society, Washington (1992).

A. Picot. *Notions générales de toxicologie moléculaire. Introduction à la toxicochimie.* ICSN, CNRS, Gif-sur-Yvette (1988).

A. Picot and J. M. Louis. *Toxicologie moléculaire: notions de biologie et de chimie appliquées.* Technique et Documentation, Lavoisier, Paris (1993).

Clinical Toxicology

M. J. Lefevre. *Urgences toxicologiques: produits chimiques industriels non pharmaceutiques.* Masson, Paris (1980).

T. A. Gossel and J. B. Douglas. *Principles of clinical toxicology.* 2nd Ed. Raven Press, New York (1989).

M. J. Ellenhorn and D. G. Barceloux. *Medical toxicology: Diagnosis and treatment of human poisoning.* Elsevier, New York (1987).

Toxicological Data

M. J. Archieri, M. J. Mageleine and A. Picot. *Les bases et banques de données accessibles en toxicologie et en hygiène et sécurité.* L'Actualité Chimique, September, p. 69–72 (1985).

P. Wexler. *Information resources in toxicology.* 2nd Ed. Elsevier, New York (1988).

———. *Registry of toxic effects of chemical substances.* NIOSH, U.S. Department of Health and Human Services, Cincinnati, Vols. 1–5 (1985–1986).

———. *Chemical information manual.* Government Institutes Inc., Rockville (1988).

Occupational Risks

General References

Ph. Lazar. *Pathologie industrielle: approche épidémiologique.* Flammarion Médecine (1979).

———. *Encyclopedia of occupational health and safety.* Vols. 1 and 2, 3rd revised Ed. International Labour Office, Geneva (1983).

B. Cassou, D. Huez, M. L. Mousel, C. Spitzer and A. Touranchet. *Les risques du travail: pour ne pas perdre sa vie à la gagner.* Editions La Découverte, Paris (1985).

———. *Encyclopédie médico-chirurgicale. Intoxications.* Médecine du travail, Editions Techniques, Paris.

A. Oudiz and D. Hemon. *Evaluation des risques et des actions de prévention en milieu professionnel. Collection INSERM Analyes et Prospectives.* La Documentation Française, Paris(1985).

———. *Face à un produit suspect, que faire?* INPACT, Paris (1986).

———. *The health and safety fact book.* Professional Publishing Ltd., London (1988).

Microbiological Risks

C. Guyot-Jeannin. *Les risques infectieux dans les laboratoires de bactériologie.* INRS no. 460, INRS, Paris (1982).

E. Hartree and V. Booth *Safety in biological laboratories*. The Biochemical Society, Colchester (1977).

———. *Laboratory safety monograph, a supplement to the NIH guidelines for recombinant DNA research*. Office of Research Safety, NIH, NCI and Special committee of safety and health experts, Bethesda (1979).

H. A. Mooney and G. Bernardi. *Introduction of genetically modified organisms into the environment*. John Wiley and Sons, Chichester (1990).

S. R. Rayburn. *The foundations of laboratory safety. A guide for the biomedical laboratory*. Springer-Verlag, New York (1990).

A. M. Ducat Man and D. Liberman. *Occupational medicine state of the art reviews: the biotechnology industry*. MIT, Cambridge (1991).

———. *La sécurité des applications industrielles des biotechnologies*. Ministère de l'Industrie, Paris (1981).

J. M. Seamer and M. Wood. *Safety in the animal house*. 2nd Ed. Laboratory Animals Ltd., London (1981).

———. *Manuel de sécurité biologique en laboratoire*. Organisation Mondiale de la Santé, Genève (1984).

M. Blanc. *L'Ere de la génétique*. Editions de la Découverte, Paris (1986).

Risks due to Radioelements

J. Rodier and J. Ph. Chassany. *Manuel de la radioprotection pratique*. Maloine, Paris (1974).

———. *La sécurité dans l'emploi des radionucléides en sources non scellées*. INRS, Paris (1978).

G. Simonnet and M. Oria. *Les mesures de radioactivité à l'aide des compteurs à scintillateur liquide*. Eyrolles, Paris (1980).

P. Galle. *Toxiques nucléaires*. Masson, Paris (1982).

D.J. Gambini and R. Granier. *Manuel pratique de radioprotection*. Technique et Documentation, Lavoisier, Paris (1992).

Risks due to Nonionizing Radiation

Rayonnement ultra-violet. Critères d'hygiène de l'environnement no. 14. Organisation Mondiale de la Santé, Genève (1980).

Fréquences radioélectriques et hyperfréquences. Critères d'hygiène de l'environnement no. 16. Organisation Mondiale de la Santé, Genève (1981).

M. J. Suess. *La protection contre les rayonnements non ionisants. Publications régionales série européenne no. 10*. Organisation Mondiale de la Santé, Copenhague (1985).

C. Polk. *CRC Handbook of biological effects of electromagnetic fields*. CRC Press, Boca Raton (1986).

J. Thuery. *Les micro-ondes et leurs effets sur la matière*. Technique et Documentation, Lavoisier, Paris (1989).

Index

Acetic acid, 128
Acetylenic derivatives, 20
Acid halides and anhydrides, destruction of, 194–196
Acids
  mineral, destruction of, 178
  organic, destruction of, 178–179
  strong, 127–128, 129
Acrylamide, 73
Acrylic derivatives, 55
Acrylonitrile, 73
Aflatoxins, destruction of, 233, 234
Aldehydes, 54, 81–82
  destruction of, 180–182
Aliphatic amines, 53
Alkali amides, destruction of, 183–184
Alkali hydrides, 26
Alkali metals, 26
Alkaline acetylides, destruction of, 173
Alkaline chlorates, 23
Alkaline earth metals, 26
Alkaline metals, 26
  destruction of, 212–213
Alkaline sulfides, destruction of, 204
Alkanes

metabolic routes for, 157
  nephrotoxic, 158–159
Alkyl halides, destruction of, 193–194
Alkyl sulfates and sulfonates, 95–97, 104–105
Alkylating agents, 42, 58–62
  mutagenicity and carcinogenicity of, 65
  principal, 64
Allergenic substances, 50–56, 57
  principal, 57
Allergies, 50
Allyl halides, destruction of, 193–194
Ames test, 69
Amides, 88
Amines, aromatic. *See* Aromatic amines
Ammonium persulfate, 24–25
Animal experiments, 247
Antimony, precipitation technique for, 211
Antitumor agents, destruction of, 234
Apparatus
  pressurized, 12
  risks associated with, 11–17
Aromatic amines, 54, 85–88, 102
  destruction of, 190–191, 234